NASA/CR-2003-212022

Asymptotic Distribution of Eigenfrequencies for a Coupled Euler-Bernoulli and Timoshenko Beam Model

Marianna A. Shubov
Texas Tech University
Lubbock, Texas

Cheryl A. Peterson
University of California, Los Angeles
Los Angeles, California

National Aeronautics and
Space Administration

Dryden Flight Research Center
Edwards, California 93523-0273

November 2003

NOTICE

Use of trade names or names of manufacturers in this document does not constitute an official endorsement of such products or manufacturers, either expressed or implied, by the National Aeronautics and Space Administration.

Available from the following:

NASA Center for AeroSpace Information (CASI)
7121 Standard Drive
Hanover, MD 21076-1320
(301) 621-0390

National Technical Information Service (NTIS)
5285 Port Royal Road
Springfield, VA 22161-2171
(703) 487-4650

Asymptotic distribution of eigenfrequencies for a coupled Euler-Bernoulli and Timoshenko beam model.

Marianna A. Shubov

Department of Mathematics and Statistics
Texas Tech University, Lubbock, TX 79409
Phone: (806)-742-2336
E-mail: marianna.shubov@ttu.edu
Fax: (806)-742-1112

Cheryl A. Peterson

Flight Systems Research Center
Department of Electrical Engineering
School of Engineering and Applied Sciences
University of California, Los Angeles, CA 90024
Phone: (310)-206-8148

CONTENTS

ABSTRACT

This research is devoted to the asymptotic and spectral analysis of a coupled Euler–Bernoulli and Timoshenko beam model. The model is governed by a system of two coupled differential equations and a two parameter family of boundary conditions modelling the action of self–straining actuators. The aforementioned equations of motion together with a two–parameter family of boundary conditions form a coupled linear hyperbolic system, which is equivalent to a single operator evolution equation in the energy space. That equation defines a semigroup of bounded operators. The dynamics generator of the semigroup is our main object of interest. For each set of boundary parameters, the dynamics generator has a compact inverse. If both boundary parameters are not purely imaginary numbers, then the dynamics generator is a nonselfadjoint operator in the energy space. We calculate the spectral asymptotics of the dynamics generator. We find that the spectrum lies in a strip parallel to the horizontal axis, and is asymptotically close to the horizontal axis – thus the system is stable, but is not uniformly stable.

CHAPTER I

Introduction

We study the spectral properties and derive the spectral asymptotics of a family of nonselfadjoint operators generated by a coupled Euler–Bernoulli and Timoshenko beam model. Such a model actually occurs in classical aeroelastic textbooks such as [1,2,3]. We formulate and prove spectral asymptotics of nonselfadjoint operators that are the dynamics generators for hyperbolic systems, which govern the motion of a coupled Euler–Bernoulli and Timoshenko beam model subject to a two–parameter family of nonconservative boundary conditions. The specific model that we consider includes a two parameter family of boundary conditions which model the action of piezo–electric materials. The precise formulation of the problem and its partial analysis are presented in paper [4].

Before we turn to the description of the model and outline the main findings in this research, we would like to emphasize the connection between the present mathematical study and the problem of flutter analysis of an aircraft wing in a subsonic air flow. The model discussed in this paper describes the so–called ground vibrations of a long and slender aircraft wing. Such a problem is the first step in the analysis of a response of an aircraft wing to the turbulent air flow when an aircraft is in–flight. The problem of ground vibrations has been known for a long time. However, an extensive mathematical and engineering literature related to the problem (often called the bending–torsion vibration model) deals either with numerical calculations or experimental verification of the numerical results. Analytical investigation of the properties of the vibrating coupled beams with nonconservative boundary conditions has been, in fact, an open problem. In the present paper, we present the first in the literature on aeroelasticity analytical formulae for the eigenfrequencies of the ground vibrations. Such an important step forward in the study of this problem has become possible not only due to the mathematical and physical backgrounds of the authors, but also due to two additional special reasons. Namely, the major part of the analysis has been accomplished when the first author (M.A. Shubov) has been awarded with Interdisciplinary Grant in the Mathematical Sciences (IGMS) by the National Science Foundation (NSF Award Number: 9972748; Amount: $100,000; Initial period: 09/01/99–08/31/00; Extension with Expiration Date: 08/31/01). Due to this award, the first author had a one–year visit to Flight Systems Research Center (FSRC) at UCLA. During this stay at FSRC, the first author was able to acquire substantial knowledge in the area of aircraft engineering while working together with the research team of the Center and by discussing different topics with the Director of FSRC, Prof. A.V. Balakrishnan. In addition to the research, done on the problem and presented in this paper, the author was able to create a basis for future research on mathematical problems arising in aircraft engineering.

The second author (C.A. Peterson) participated in this study as a graduate student of Dr. M.A. Shubov. For her research, the second author has been awarded with two Texas Space Grant Consortium Fellowships (in 2000–2001 and in 2001–2002 school years). At the school year 2002–2003, the second author is a Post–graduate Research Engineer at FSRC (UCLA), where she is working on numerical simulations to verify the efficiency of the spectral asymptotics obtained analytically in the present paper.

Now we briefly describe the organization of the paper.

• In Chapter I, we introduce a coupled system of differential equations, which will be our main system in the present work. We also introduce a two–parameter family of boundary conditions, which contains *two control parameters* (see also [5, 6]).

• In Chapter II, we justify the new setting of the initial boundary–value problem in the form of the first order in time *evolution equation*. The asymptotic properties of the spectrum of the dynamics generator is our main interest in the work.

• In Chapter III, we initiate methodical study of the spectrum of the dynamics generator. As the first step, we reduce the problem for the spectrum of the generator to the problem of finding spectral asymptotics for the corresponding *operator pencil*. We recall the definitions related to operator pencils and provide the explanation why the pencil considered in the present work is highly nonstandard and extremely complicated. We show that the first necessary step is to analyze the asymptotic behavior of the fundamental system of solutions of the sixth order ordinary differential equation involving the spectral parameter λ (see Eq.(3.16) below). The latter analysis requires us to derive asymptotics for the roots of the sixth order polynomial of a special type.

• In Chapter IV, we derive asymptotic approximations for the roots of the characteristic polynomial, which is associated to the aforementioned sixth order ordinary differential equation. Namely, we obtain the approximations for six roots of the sixth order polynomial, approximations when the spectral parameter $|\lambda| \to \infty$ and those approximations are *uniform with respect to the spatial variable x*. Technically, this Chapter is very complicated and the results obtained in it are of crucial importance for the remaining Chapters V–VIII.

• In Chapter V, we introduce a special method to solve the boundary problem, i.e., we use the so–called *two–step procedure*. This is a relatively new method that has been introduced in papers [7–9] to solve the boundary–value problem for a spatially nonhomogeneous Timoshenko beam model. By using the aforementioned procedure, we examine the effect of applying the boundary conditions from the left and from the right separately. Namely, in Section 5.2, we look for a solution of the spectral equation for the operator pencil, a solution which satisfies only three left–hand side boundary conditions without any restrictions on the behavior of such a solution at the right–hand side of the flexible structure. In this Section, we introduce an important notion which we call the *left–reflection matrix* and denote it as $\mathbb{R}_l$.

• In Chapter VI, we derive an asymptotic approximation for the *right–reflection matrix*, which is similar to the left–reflection matrix of Chapter V. The right reflection matrix is useful to describe the solution of the main differential equation, which satisfies only the right–hand side boundary conditions without imposing any restrictions on the solution at the left end.

• In Chapter VII, we incorporate all information obtained in the previous two Chapters to derive the spectral asymptotics. The main tool in proving asymptotical for-

3

mulae (3.2)–(3.3) with the necessary accuracy is the well–known *Rouche's Theorem*.

Euler–Bernoulli/Timoshenko Beam Model

From now on, we will focus on the boundary–value problem consisting of a system of two coupled hyperbolic partial differential equations in two unknown functions h and α, with $h(t,x)$ being a deflection at a point x at a time moment t and $\alpha(t,x)$ being a torsion angle at a point x and a time t. We assume that the spatial extent of the flexible structure is L and $t > 0$.

$$
\begin{cases}
m\ddot{h}(t,x) + S\ddot{\alpha}(t,x) + EIh''''(t,x) = 0, & -L < x < 0; \quad 0 < t, \\
\\
S\ddot{h}(t,x) + I_\alpha\ddot{\alpha}(t,x) - GJ\alpha''(t,x) = 0, & -L < x < 0; \quad 0 < t,
\end{cases}
\tag{1.1}
$$

where m is a mass per unit length, S is a mass moment per unit length, EI is a bending stiffness, GJ is a torsion stiffness, and I_α is a moment of inertia. The system is supplied with a two–parameter family of boundary conditions

$$
h(t,-L) = h'(t,-L) = 0 \quad \alpha(t,-L) = 0;
\tag{1.2}
$$

$$
\begin{aligned}
h'''(t,0) &= 0, \\
EIh''(t,0) + g_h\dot{h}'(t,0) &= 0, \\
GJ\alpha'(t,0) + g_\alpha\dot{\alpha}(t,0) &= 0.
\end{aligned}
\tag{1.3}
$$

In (1.1)–(1.3), we have denoted the time derivative by the over dot.

We notice now that the set of boundary conditions at the left end is standard. However, the right–hand side boundary conditions are highly nonstandard, i.e., they contain two arbitrary parameters g_h and g_α. These parameters are used in current mathematical and engineering literature in order to model the action of "smart materials" (see [10–13]). Technically, the analysis can be carried out for any complex values of the boundary parameters. An important case for practical applications is

4

the one when g_h and g_α are positive numbers. In this case, g_h is called *a bending control gain* and g_α is called *a torsion control gain*.

In addition to the boundary conditions, we introduce a set of initial conditions in a standard manner

$$h(0, x) = h_0(x), \ \dot{h}(t, x)\big|_{t=0} = h_1(x), \ \alpha(0, t) = \alpha_0(x), \ \dot{\alpha}(t, x)\big|_{t=0} = \alpha_1(x). \qquad (1.4)$$

To conclude the Introduction, we note that we consider a beam that is perfectly elastic, and rigid in cross sections perpendicular to the lengthwise direction. It has an elastic axis, implying that the wing is unswept and without structural discontinuities, so that elastic coupling between bending and twisting is eliminated. The elastic axis is straight, and rotary inertia and shear deformation is neglected.

CHAPTER II

Dynamics Generator and Its General Spectral Properties

2.1 Dynamics Generator

To analyze the initial boundary–value problem (1.1)–(1.3), we will present it as a first order in time evolution equation. *The dynamics generator* will be our main object of interest. We will show that the dynamics generator is a 4×4 matrix differential operator, which acts in the so–called energy space. To introduce the formula for the dynamics generator and to describe its domain, we carry out the following steps. First, we can verify that our system of equations

$$\begin{cases} m\ddot{h}(t,x) + S\ddot{\alpha}(t,x) = -EIh''''(t,x), \\ S\ddot{h}(t,x) + I_\alpha\ddot{\alpha}(t,x) = GJ\alpha''(t,x). \end{cases} \tag{2.1}$$

can be represented in the form

$$\begin{bmatrix} 1 & 0 & 0 & 0 \\ 0 & 1 & 0 & 0 \\ 0 & 0 & m & S \\ 0 & 0 & S & I_\alpha \end{bmatrix} \begin{bmatrix} \dot{h}(t,x) \\ \dot{\alpha}(t,x) \\ \ddot{h}(t,x) \\ \ddot{\alpha}(t,x) \end{bmatrix} = \begin{bmatrix} 0 & 0 & 1 & 0 \\ 0 & 0 & 0 & 1 \\ -EI\dfrac{\partial^4}{\partial x^4} & 0 & 0 & 0 \\ 0 & GJ\dfrac{\partial^2}{\partial x^2} & 0 & 0 \end{bmatrix} \begin{bmatrix} h(t,x) \\ \alpha(t,x) \\ \dot{h}(t,x) \\ \dot{\alpha}(t,x) \end{bmatrix}. \tag{2.2}$$

If we introduce a 4–component vector Y by the formula $Y = (h, \alpha, \dot{h}, \dot{\alpha})^T$ (the superscript "T" means transposition), and denote the matrices in (2.2) by M and A, then Eq.(2.2) can be written in the form

$$\mathbf{M}\dot{Y} = \mathbf{A}Y. \tag{2.3}$$

Now we introduce an important assumption

$$\Delta = mI_\alpha - S^2 > 0. \tag{2.4}$$

Due to (2.4), we can rewrite Eq.(2.3) as

$$\dot{Y} = \mathbf{M}^{-1}\mathbf{A}Y = i[-i\mathbf{M}^{-1}\mathbf{A}]Y. \tag{2.5}$$

Denoting $\mathcal{L} = -i\mathbf{M}^{-1}\mathbf{A}$, we finally rewrite Eq.(2.5) in the desired form

$$\dot{Y} = i\mathcal{L}Y. \tag{2.6}$$

If $\Phi^T(t,x) = \{\phi_0(t,x), \phi_1(t,x), \phi_2(t,x), \phi_3(t,x)\} = \{h, \alpha, \dot{h}, \dot{\alpha}\}$, $-L \leq x \leq 0$, $t \geq 0$, then the initial–boundary value problem (1.1)–(1.3) can be rewritten in the form of the first order in time evolution equation

$$\dot{\Phi} = i\mathcal{L}\Phi, \quad \Phi\big|_{t=0} = \Phi_0, \tag{2.7}$$

with the dynamics generator $\mathcal{L}$ being defined on smooth functions $\Phi = (\phi_0, \phi_1, \phi_2, \phi_3)^T$ by the formula

$$\mathcal{L} = -i \begin{bmatrix} 0 & 0 & 1 & 0 \\ 0 & 0 & 0 & 1 \\ \dfrac{-I_\alpha EI}{\Delta}\dfrac{\partial^4}{\partial x^4} & \dfrac{-SGJ}{\Delta}\dfrac{\partial^2}{\partial x^2} & 0 & 0 \\ \dfrac{S\,EI}{\Delta}\dfrac{\partial^4}{\partial x^4} & \dfrac{mGJ}{\Delta}\dfrac{\partial^2}{\partial x^2} & 0 & 0 \end{bmatrix}. \tag{2.8}$$

subject to the following boundary conditions:

$$\phi_0(-L) = \phi_0'(-L) = \phi_1(-L) = 0,,$$
$$\phi_0'''(0) = 0,$$
$$EI\phi_0''(0) + g_h\phi_2'(0) = 0,$$
$$GJ\phi_1'(0) + g_\alpha\phi_3(0) = 0. \tag{2.9}$$

2.2 Operator Setting

Let us consider the energy of the system by carrying out the following steps. Let us multiply the first equation from system (2.1) by $\bar{h}_t$ and the second equation from system (2.1) by $\bar{\alpha}_t$ and then add the resulting two equations. If we denote the resulting sum as EQ1, we have

$$EQ1 = mh_{tt}(t,x)\bar{h}_t(t,x) + S\alpha_{tt}(t,x)\bar{h}_t(t,x) + EI\,h''''(t,x)\bar{h}_t(t,x)+$$
$$Sh_{tt}(t,x)\bar{\alpha}_t(t,x) + I_\alpha\alpha_{tt}(t,x)\bar{\alpha}_t(t,x) - GJ\,\alpha''(t,x)\bar{\alpha}_t(t,x) = 0. \tag{2.10}$$

7

Let us take the complex conjugate of Eq.(2.10) and denote this equation by EQ2. It can be verified by a direct calculation that we have the following result:

$$EQ1 + EQ2 = m\frac{d}{dt}|h_t(t,x)|^2 + S\frac{d}{dt}\left(\alpha_t(t,x)\bar{h}_t(t,x) + \bar{\alpha}_t(t,x)h_t(t,x)\right) +$$

$$EI\left(h''''(t,x)\bar{h}_t(t,x) + \bar{h}''''(t,x)h_t(t,x)\right) + I_\alpha\frac{d}{dt}|\alpha_t(t,x)|^2 - \quad (2.11)$$

$$GJ\left(\alpha''(t,x)\bar{\alpha}_t(t,x) + \bar{\alpha}''(t,x)\alpha_t(t,x)\right) = 0.$$

Eq.(2.11) suggests that a convenient expression for the energy of the system can be taken in the form of the following functional:

$$\mathcal{E}(t) = \frac{1}{2}\int_{-L}^{0}\Big[EI\,|h''(t,x)|^2 + GJ\,|\alpha'(t,x)|^2 + m\,|h_t(t,x)|^2 +$$

$$I_\alpha\,|\alpha_t(t,x)|^2 + S\left(\alpha_t(t,x)\bar{h}_t(t,x) + \bar{\alpha}_t(t,x)h_t(t,x)\right)\Big]dx. \quad (2.12)$$

The following Lemma regarding this energy has been proved in our paper [14].

Lemma 2.1. *Under condition (2.4), the energy of vibrations, given by formula (2.12), is nonnegative and is equal to zero if and only if* $h(t,x) = \alpha(t,x) = 0$, $x \in [-L, 0]$, $t \geq 0$.

With this energy of vibrations, we can define the operator setting of the problem. First we describe the state space of the system, which will be denoted by $\mathcal{H}$. Let $\mathcal{H}$ be a set of 4–component vector–valued functions $\Phi = (\phi_0, \phi_1, \phi_2, \phi_3)^T$ obtained as a closure of smooth functions satisfying the conditions

$$\phi_0(-L) = \phi_0'(-L) = \phi_1(-L) = 0 \quad (2.13)$$

in the following energy norm:

$$\|\Phi\|_{\mathcal{H}}^2 = \frac{1}{2}\int_{-L}^{0}\Big[EI\,|\phi_0''(x)|^2 + GJ\,|\phi_1'(x)|^2 + m\,|\phi_2(x)|^2 + I_\alpha\,|\phi_3(x)|^2 +$$

$$S\left(\bar{\phi}_2(x)\phi_3(x) + \phi_2(x)\bar{\phi}_3(x)\right)\Big]dx. \quad (2.14)$$

This is shown to be a norm in our paper [14]. The operator $\mathcal{L}$ is given by formula

(2.8) and is defined on the domain

$$D\left(\mathcal{L}\right) = \Big\{ \Phi \in \mathcal{H} : \phi_0 \in H^4(-L, 0), \ \phi_1 \in H^2(-L, 0), \ \phi_2 \in H^2(-L, 0), \ \phi_3 \in H^1(-L, 0);$$

$$\phi_0(-L) = \phi_0'(-L) = \phi_1(-L) = 0, \ \phi_0'''(0) = 0;$$

$$EI \ \phi_0''(0) + g_h \phi_2'(0) = 0, \ GJ \ \phi_1'(0) + g_\alpha \phi_3(0) = 0 \Big\},$$

$$(2.15)$$

where H^i, $i = 1, 2, 4$, are the standard Sobolev spaces [15].

2.3 Properties of the Dynamics Generator

The following Lemma is proved in our paper [14].

Theorem 2.1. *The operator $\mathcal{L}$ has the following properties.*

(i) $\mathcal{L}$ is an unbounded, closed, nonselfadjoint (unless $\Re g_h = \Re g_\alpha = 0$) operator in $\mathcal{H}$.

(ii) If $\Re g_h \geq 0$ and $\Re g_\alpha \geq 0$, then $\mathcal{L}$ is a dissipative operator in $\mathcal{H}$ (i.e., if $\Phi \in D(\mathcal{L})$, then $\Im (\mathcal{L}\Phi, \Phi)_{\mathcal{H}} \geq 0$, [16]).

(iii) The inverse operator $\mathcal{L}^{-1}$ exists and it is a compact operator in $\mathcal{H}$. Therefore, $\mathcal{L}^{-1}$ has a purely discrete spectrum consisting of normal eigenvalues. (Recall that λ is a normal eigenvalue of a bounded operator A in the space H if a) λ is an isolated point of the spectrum of A, b) the algebraic multiplicity of λ is finite, c) the range $(A - \lambda I) H$ of the operator $(A - \lambda I)$ is closed [17, 18]).

We emphasize that from Theorem 2.1, the following two important results can be seen immediately: (a) the operator $\mathcal{L}$ has a purely discrete spectrum, which can accumulate only at infinity, and (b) when $\Re g_h \geq 0$ and $\Re g_\alpha \geq 0$, the spectrum is located in the closed upper half–plane.

Remark 2.1. As is accustomed in the engineering literature, we reformulate the conclusions of Theorem 2.1 for the operator $\mathfrak{L} = i\mathcal{L}$. For $\mathfrak{L}$, we obtain that when $\Re g_h \geq 0$ and $\Re g_\alpha \geq 0$, then the spectrum of this operator is located in the closed left half–plane, and consists of, at most, a countable number of eigenvalues that can accumulate only at infinity.

9

CHAPTER III

General Solution of Spectral Equation

3.1 Precise Statement of the Asymptotics of the Spectrum.

We formulate now a precise statement of the spectral results, which will be proven in the rest of the paper.

Theorem 3.1. *(a) The operator $\mathcal{L}$ has a countable set of complex eigenvalues. Under the assumption*

$$g_\alpha \neq \sqrt{I_\alpha\, GI}, \tag{3.1}$$

the set of eigenvalues is located in a strip parallel to the real axis.

(b) The entire set of the eigenvalues asymptotically splits into two disjoint subsets. We call them the h–branch and the α–branch and denote these branches by $\{\lambda_n^h\}_{n\in\mathbb{Z}}$ and $\{\lambda_n^\alpha\}_{n\in\mathbb{Z}}$ respectively. If $\Re\, g_\alpha \geq 0$ and $\Re\, g_h > 0$, then the α–branch is asymptotically close to some horizontal line in the upper half–plane. If $\Re\, g_h = \Re\, g_\alpha = 0$, then the operator $\mathcal{L}$ is selfadjoint and thus its spectrum is real. The entire set of eigenvalues may have only two points of accumulation: $+\infty$ and $-\infty$ in the sense that $\Re\, \lambda_n^{h(\alpha)} \to \pm\infty$ and $\Im\, \lambda_n^{h(\alpha)} < const$ as $n \to \pm\infty$ (see formulae (3.2) and (3.3) below).

(c) The following asymptotic formula is valid for the h–branch of the spectrum:

$$\lambda_n^h = \pm\pi^2/L^2\sqrt{I_\alpha EI/\Delta}(|n| - 1/4)^2 + O(1), \quad |n| \to \infty. \tag{3.2}$$

In formula (3.2), the sign "+" should be taken for $n > 0$ and "−" for $n < 0$.

(d) The following asymptotic formula is valid for the α–branch of the spectrum:

$$\lambda_n^\alpha = \frac{\pi n}{L\sqrt{I_\alpha/GJ}} + \frac{i}{2L\sqrt{I_\alpha/GJ}} \ln \frac{g_\alpha + \sqrt{I_\alpha\, GJ}}{g_\alpha - \sqrt{I_\alpha\, GJ}} + O(|n|^{-1/2}), \quad |n| \to \infty. \tag{3.3}$$

In (3.3), the principle value of the logarithm is understood.

3.2 Operator Pencil

In this section, we introduce an operator–valued polynomial function, which we call an *operator pencil*. To introduce this operator pencil, which is associated to the dynamics generator $\mathcal{L}$, we start with the spectral equation for this operator

$$\mathcal{L}\Phi = \lambda\Phi, \quad \Phi \in \mathcal{D}(\mathcal{L}). \tag{3.4}$$

Using the explicit formula for $\mathcal{L}$, we obtain

$$-i\begin{bmatrix} 0 & 0 & 1 & 0 \\ 0 & 0 & 0 & 1 \\ \dfrac{-I_\alpha EI}{\Delta}\dfrac{\partial^4}{\partial x^4} & \dfrac{-SGJ}{\Delta}\dfrac{\partial^2}{\partial x^2} & 0 & 0 \\ \dfrac{SEI}{\Delta}\dfrac{\partial^4}{\partial x^4} & \dfrac{mGJ}{\Delta}\dfrac{\partial^2}{\partial x^2} & 0 & 0 \end{bmatrix}\begin{bmatrix} \phi_0 \\ \phi_1 \\ \phi_2 \\ \phi_3 \end{bmatrix} = \lambda\begin{bmatrix} \phi_0 \\ \phi_1 \\ \phi_2 \\ \phi_3 \end{bmatrix}. \tag{3.5}$$

Rewriting Eq.(3.5) component–wise yields the following four equations:

$$\phi_2 = i\lambda\phi_0, \tag{3.6}$$

$$\phi_3 = i\lambda\phi_1, \tag{3.7}$$

$$\frac{I_\alpha EI}{\Delta}\phi_0^{IV} + \frac{SGJ}{\Delta}\phi_1'' = -i\lambda\phi_2, \tag{3.8}$$

$$\frac{SEI}{\Delta}\phi_0^{IV} + \frac{mGJ}{\Delta}\phi_1'' = i\lambda\phi_3. \tag{3.9}$$

Our goal is to eliminate the three components ϕ_1, ϕ_2, and ϕ_3 from system (3.6)–(3.9) and to derive a single equation with respect to the one component ϕ_0. Substituting Eqs.(3.6) and (3.7) into Eqs.(3.8) and (3.9) to eliminate ϕ_2 and ϕ_3, we obtain

$$I_\alpha EI\,\phi_0^{IV} + SGJ\,\phi_1'' = \Delta\lambda^2\phi_0, \tag{3.10}$$

$$SEI\phi_0^{IV} + mGJ\phi_1'' = -\Delta\lambda^2\phi_1, \tag{3.11}$$

where Δ is defined in (2.4). We will now eliminate ϕ_1 from Eqs. (3.10) and (3.11). To this end, let us solve Eq.(3.10) for ϕ_1'' and obtain

$$\phi_1'' = \frac{1}{SGJ}[-I_\alpha EI\phi_0^{IV} + \lambda^2\Delta\phi_0]. \tag{3.12}$$

11

Differentiating both sides of the latter equation gives

$$\phi_1^{IV} = \frac{1}{SGJ}[-I_\alpha EI\phi_0^{VI} + \lambda^2\Delta\phi_0''].$$ (3.13)

By twice differentiating both sides of Eq.(3.11), we obtain

$$S\,EI\phi_0^{VI} + mGJ\phi_1^{IV} = -\lambda^2\Delta\phi_1''.$$ (3.14)

Substituting (3.12) and (3.13) into Eq.(3.14) and multiplying both sides by SGJ gives

$$S^2\,GJ\,EI\phi_0^{VI} + mGJ[-I_\alpha EI\phi_0^{VI} + \lambda^2\Delta\phi_0''] = \lambda^2\Delta[I_\alpha EI\phi_0^{IV} - \lambda^2\Delta\phi_0].$$ (3.15)

Collecting like terms in Eq.(3.15) and then simplifying, we arrive at the following final form of the equation for the component ϕ_0:

$$EI\,GJ\phi_0^{VI} + \lambda^2 I_\alpha EI\phi_0^{IV} - \lambda^2 m\,GJ\phi_0'' - \lambda^4\Delta\phi_0 = 0.$$ (3.16)

Thus, we have defined an equation, in which ϕ_0 and its derivatives are the only unknown functions. We notice that the left–hand side of Eq.(3.16) is a fourth order polynomial with respect to λ. However, coefficients in that polynomial are high order differential operations. It is convenient to introduce special notation for this operation.

Let $\mathcal{P}(\cdot)$ be a polynomial operation defined by the formula

$$\mathcal{P}(\lambda)\phi_0 = EI\,GJ\phi_0^{VI} + \lambda^2 I_\alpha EI\phi_0'''' - \lambda^2 mGJ\phi_0'' - \lambda^4\Delta\phi_0,$$ (3.17)

where ϕ_0 is a smooth function for $x \in [-L, 0]$. In order to determine the domain of $\mathcal{P}(\cdot)$, we have to calculate the conditions, which ϕ_0 inherits from the domain of the operator $\mathcal{L}$. To do this, we will take the boundary conditions from the domain of the dynamics generator $\mathcal{L}$ (see 2.9) and rewrite them in terms of ϕ_0 and its derivatives. First of all, we need to write ϕ_1 in terms of ϕ_0. To carry out this elimination, we return to the system of two equations (3.10) and (3.11). From this system of two equations, we will eliminate ϕ_1'' to obtain one equation involving ϕ_0 and its derivatives and ϕ_1

12

which will then be solved for. We will accomplish the following sequence of steps: first, we divide Eq.(3.10) by S and Eq.(3.11) by m; secondly, we subtract the second equation from the first one and have

$$\phi_0^{IV} \left[\frac{I_\alpha \, EI}{S} - \frac{S \, EI}{m} \right] = \frac{\Delta \lambda^2}{S} \phi_0 + \frac{\Delta \lambda^2}{m} \phi_1. \tag{3.18}$$

Simplifying Eq.(3.18) and taking into account formula (2.4) for Δ, we obtain the representation for ϕ_1 in terms of ϕ_0

$$\phi_1 = \frac{EI}{S} \lambda^{-2} \phi_0^{IV} - \frac{m}{S} \phi_0. \tag{3.19}$$

This expression for ϕ_1 will be substituted along with (3.6) and (3.7), into the boundary conditions (2.9) to obtain precise boundary conditions needed for the function ϕ_0 to be in the domain of the operator pencil.

The first two boundary conditions, being only in terms of ϕ_0, remain unchanged, i.e., we have $\phi_0(-L) = 0$ and $\phi_0'(-L) = 0$. Substituting (3.19) into the third boundary condition yields

$$EI \phi_0^{IV}(-L) - \lambda^2 m \phi_0(-L) = 0. \tag{3.20}$$

So far, we have determined the left–hand boundary conditions. Now we derive appropriate forms for the right–hand boundary conditions. The fourth boundary condition of (2.9), being only in terms of ϕ_0, remains the same, i.e. $\phi_0'''(0) = 0$. The fifth boundary condition of (2.9), after substitution of (3.6), becomes

$$EI \phi_0''(0) + i\lambda g_h \phi_0'(0) = 0. \tag{3.21}$$

The only boundary condition that remains to be determined is the sixth one. This condition after substitution of (3.7) becomes

$$GJ \phi_1'(0) + i\lambda g_\alpha \phi_1(0) = 0. \tag{3.22}$$

Replacing ϕ_1 according to formula (3.19) leads to

$$GJ \left[\frac{EI}{S} \lambda^{-2} \phi_0^{V}(0) - \frac{m}{S} \phi_0'(0) \right] + g_\alpha i\lambda \left[\frac{EI}{S} \lambda^{-2} \phi_0^{IV}(0) - \frac{m}{S} \phi_0(0) \right] = 0. \tag{3.23}$$

13

Multiplying Eq.(3.23) by $S\lambda^2$ and rearranging its terms, we arrive at the following boundary condition:

$$EI\,GJ\,\phi_0^V(0) + i\lambda EI\,g_\alpha\phi_0^{IV}(0) - \lambda^2 m\,GJ\,\phi_0'(0) - i\lambda^3 g_\alpha m\phi_0(0) = 0. \qquad (3.24)$$

Therefore, the problem of finding the eigenvalues and eigenfunctions of the operator $\mathcal{L}$ (see Eq.(3.5)) has been reduced to the problem of finding those values of the parameter λ for which the sixth order ordinary differential equation (3.16) has nontrivial solutions satisfying six boundary conditions.

Now we are in a position to introduce a pencil associated with the operator $\mathcal{L}$. We recall [19] that a *polynomial operator pencil* $A(\lambda)$ is an operator–valued function defined by the formula $A(\lambda) = \lambda^n + \lambda^{n-1}A_{n-1} + ... + A_0$ in which A_k are linear operators. Those operators may be either bounded or unbounded and either selfadjoint or nonselfadjoint. The degree n of this polynomial is called the order of the pencil. Let $\mathcal{P}(\cdot)$ be the fourth order operator pencil that acts on a function $\phi \in H^6(-L,0)$ by the formula

$$\mathcal{P}(\lambda)\phi = EI\,GJ\phi^{VI} + \lambda^2 I_\alpha EI\phi^{IV} - \lambda^2 m\,GJ\phi'' - \lambda^4 \Delta\phi \qquad (3.25)$$

and is defined on the domain

$$D(\mathcal{P}) = \{\phi \in H^6(-L,0) : \phi(-L) = \phi'(-L) = \phi^{IV}(-L) = 0;$$

$$EI\phi''(0) + g_h i\lambda\phi'(0) = 0, \quad \phi'''(0) = 0, \qquad (3.26)$$

$$EI\,GJ\,\phi^V(0) + EI\,g_\alpha i\lambda\phi^{IV}(0) - m\,GJ\,\lambda^2\phi'(0) - g_\alpha m i\lambda^3\phi(0) = 0\}.$$

We note that H^6 is the standard Sobolev space [15]. We call a nontrivial solution $\phi \in D(\mathcal{P})$ of the pencil equation $\mathcal{P}(\lambda)\phi = 0$ an *eigenfunction* of the pencil $\mathcal{P}(\cdot)$ and the corresponding value of λ an *eigenvalue* of $\mathcal{P}(\cdot)$. It is clear that having an eigenfunction of the pencil and using (3.19), we can find ϕ_1 and then find all four components of the eigenvector of the operator $\mathcal{L}$.

We mention that $\mathcal{P}(\cdot)$ is a nonstandard pencil due to the fact that the spectral parameter λ enters the domain explicitly. These type of pencils have not been considered in the monograph [19]. However, it is convenient to keep the terminology because there exists an extensive literature in which the pencils with the parameter dependent boundary conditions appear naturally.

3.3 Characteristic Equation

In this section, we initiate the analysis of the pencil equation $\mathcal{P}(\lambda)\phi = 0$. In particular, in the present section we focus on the differential equation (3.16), which is a sixth order ordinary differential equation with constant coefficients containing the complex parameter λ. We are looking for the asymptotic representation for the fundamental system of its solutions. It is important to mention that we are looking for the asymptotics with respect to λ when $|\lambda| \to \infty$ and those asymptotics must be uniform with respect to the spatial variable $x \in [-L, 0]$.

As is well–known, to find the fundamental system of solutions of the sixth order ordinary differential equation, we have to find appropriate approximations for the roots of the sixth order polynomial, which is the characteristic polynomial for the differential equation. The characteristic equation has the following form:

$$EI\,GJ(x^2)^3 + \lambda^2 I_\alpha\,EI(x^2)^2 - \lambda^2 m\,GJ(x^2) - \Delta\lambda^4 = 0. \tag{3.27}$$

Clearly, Eq.(3.27) is of a sixth order, but if we change an independent variable, we can reduce it to a cubic equation. Namely, let

$$y = x^2, \tag{3.28}$$

and then rewriting Eq.(3.27) in terms of y, we have

$$EI\,GJy^3 + \lambda^2 I_\alpha\,EIy^2 - \lambda^2 m\,GJy - \Delta\lambda^4 = 0. \tag{3.29}$$

Using Cardano's Formulae [20], we can obtain the solution of a cubic polynomial. However, in order to apply those formulae, we need the cubic polynomial to be monic with no quadratic term. Let us reduce the polynomial in (3.29) to the desired form. Dividing both sides of (3.29) by $EI\,GJ$ yields

$$y^3 + \left(\frac{I_\alpha\lambda^2}{GJ}\right)y^2 - \left(\frac{m\lambda^2}{EI}\right)y - \frac{\Delta\lambda^4}{EI\,GJ} = 0, \tag{3.30}$$

which is a monic polynomial. Next we need to make the quadratic term in (3.30) vanish. To do so, we notice that for a polynomial such as

$$f(x) = x^3 + a_2 x^2 + a_1 x + a_0, \tag{3.31}$$

the substitution

$$x = z - a_2/3 \tag{3.32}$$

will result in a polynomial in z, with no quadratic term [20]. So if we substitute

$$y = z - \frac{1}{3}\left(\frac{I_\alpha}{GJ}\lambda^2\right) \tag{3.33}$$

into Eq.(3.30), then we have an equation adjusted to Cardano's Formulae. Taking into account that

$$y^3 = y(y^2) = \left(z - \frac{I_\alpha \lambda^2}{3GJ}\right)\left(z^2 - \frac{2I_\alpha \lambda^2}{3GJ}z + \left(\frac{I_\alpha \lambda^2}{3GJ}\right)^2\right)$$
$$= z^3 - \frac{I_\alpha \lambda^2}{GJ}z^2 + \frac{1}{3}\left(\frac{I_\alpha \lambda^2}{GJ}\right)^2 z - \left(\frac{I_\alpha \lambda^2}{3GJ}\right)^3, \tag{3.34}$$

and substituting (3.33) and (3.34) into Eq.(3.30), we obtain

$$\left[z^3 - \frac{I_\alpha \lambda^2}{GJ}z^2 + \frac{1}{3}\left(\frac{I_\alpha \lambda^2}{GJ}\right)^2 z - \left(\frac{I_\alpha \lambda^2}{3GJ}\right)^3\right]\frac{I_\alpha \lambda^2}{GJ}\left[z^2 - \frac{2I_\alpha \lambda^2}{3GJ}z + \left(\frac{I_\alpha \lambda^2}{3GJ}\right)^2\right] -$$

$$\left(\frac{m\lambda^2}{EI}\right)\left[z - \frac{1}{3}\left(\frac{I_\alpha \lambda^2}{GJ}\right)\right] - \frac{\Delta \lambda^4}{EI\,GJ} = 0, \tag{3.35}$$

and cancel the two quadratic terms as expected. Then we combine same powers of z and have

$$z^3 + \left[\frac{1}{3}\left(\frac{I_\alpha \lambda^2}{GJ}\right)^2 - \frac{2}{3}\left(\frac{I_\alpha \lambda^2}{GJ}\right)^2 - \frac{m\lambda^2}{EI}\right]z+ \tag{3.36}$$

$$\left[-\left(\frac{I_\alpha \lambda^2}{3GJ}\right)^3 + \frac{1}{3^2}\left(\frac{I_\alpha \lambda^2}{GJ}\right)^3 + \frac{1}{3}\left(\frac{m\lambda^2}{EI}\right)\left(\frac{I_\alpha \lambda^2}{GJ}\right) - \frac{\Delta \lambda^4}{EI\,GJ}\right] = 0.$$

Simplifying the coefficients in Eq.(3.36), we finally obtain the monic cubic polynomial with no quadratic term as

$$z^3 - \left[\frac{1}{3}\left(\frac{I_\alpha \lambda^2}{GJ}\right)^2 + \frac{m\lambda^2}{EI}\right]z + \left[2\left(\frac{I_\alpha \lambda^2}{3GJ}\right)^3 + \frac{I_\alpha m - 3\Delta}{3EI\,GJ}\lambda^4\right] = 0. \qquad (3.37)$$

It is this polynomial that we will apply Cardano's Formulae to in the next section.

We recall that our goal is to find asymptotic approximations for six roots of Eq.(3.27) when $|\lambda| \to \infty$. However, before proceeding to find the six roots, we will see what knowledge may be gained by investigating the asymptotic behavior of the solutions of a simpler equation than Eq.(3.37). We will call this simpler equation the *model equation*. Let us rewrite Eq.(3.37) in the asymptotical form as $|\lambda| \to \infty$.

$$z^3 - \frac{\lambda^4}{3}\left(\frac{I_\alpha}{GJ}\right)^2(1 + O(\lambda^{-2}))z + \lambda^6 \frac{2I_\alpha^3}{3^3(GJ)^3}(1 + O(\lambda^{-2})) = 0. \qquad (3.38)$$

By omitting the lower order terms $O(\lambda^{-2})$ in Eq.(3.35), we obtain the model equation

$$z^3 - \frac{\lambda^4}{3}\left(\frac{I_\alpha}{GJ}\right)^2 z + \lambda^6 \frac{2I_\alpha^3}{3^3(GJ)^3} = 0. \qquad (3.39)$$

Assuming that a solution z_1 will be a multiple of λ^2, we substitute

$$z_1 = a\lambda^2 \qquad (3.40)$$

into Eq.(3.39) and divide by λ^6 to obtain a cubic equation for the multiple a

$$a^3 - \left(\frac{I_\alpha}{GJ}\right)^2 \frac{1}{3}a + \frac{2}{3^3}\left(\frac{I_\alpha}{GJ}\right)^3 = 0, \qquad (3.41)$$

from which we guess that a solution a will be a multiple of I_α/GJ. So, now making the substitution $a = bI_\alpha/GJ$ and then dividing by $(I_\alpha/GJ)^3$ yields

$$b^3 - \frac{1}{3}b + \frac{2}{3^3} = 0. \qquad (3.42)$$

One can check directly that a solution of this equation is $b = 1/3$. Thus we find that one solution z_1 of the model equation (3.39) can be found by successive substitutions into (3.40) as follows:

$$z_1 = a\lambda^2 = \frac{I_\alpha \lambda^2}{3GJ} = \lambda^2 \frac{R}{3}, \qquad (3.43)$$

where

$$R = \frac{I_\alpha}{GJ}.$$ (3.44)

Factoring the model equation yields

$$\left(z - \frac{I_\alpha \lambda^2}{3GJ} \right) \left(z^2 + \frac{I_\alpha \lambda^2}{3GJ} z - \frac{2}{9} \left(\frac{I_\alpha}{GJ} \right)^2 \lambda^4 \right) = 0.$$ (3.45)

The roots of the quadratic polynomial (3.45) are

$$z_2 = \frac{1}{3} \left(\frac{I_\alpha}{GJ} \right) \lambda^2, \qquad z_3 = -\frac{2}{3} \left(\frac{I_\alpha}{GJ} \right) \lambda^2.$$ (3.46)

Thus we have the following three solutions to the model equation:

$$z_1 = z_2 = \frac{I_\alpha \lambda^2}{3GJ}, \qquad z_3 = -\frac{2}{3} \left(\frac{I_\alpha}{GJ} \right) \lambda^2.$$ (3.47)

Note that $z_1 = z_2$. From the latter fact, we can expect that two roots of Eq.(3.37) will have a similar behavior in nature, while the third solution will behave differently. This exact difference in behavior remains to be seen.

CHAPTER IV

Asymptotic Analysis of the Roots of the Characteristic Polynomial

4.1 Cardano's Formulae

It is well–known, Cardano's Formulae [20] give a solution for a monic cubic equation with no quadratic term such as

$$z^3 + pz + q = 0, \tag{4.1}$$

where p and q are constants. The solution given by Cardano's Formulae to Eq.(4.1) can be represented in the form

$$z = u - v, \tag{4.2}$$

where

$$u = \sqrt[3]{-\frac{q}{2} + \sqrt{\left(\frac{q}{2}\right)^2 + \left(\frac{p}{3}\right)^3}}, \quad v = \sqrt[3]{\frac{q}{2} + \sqrt{\left(\frac{q}{2}\right)^2 + \left(\frac{p}{3}\right)^3}}. \tag{4.3}$$

We will use formulae (4.2) and (4.3) to find solutions to the characteristic equation (3.27). Recall that we made substitution (3.28) that

$$x^2 = y, \tag{4.4}$$

to obtain a cubic polynomial. Next we substituted the shift of (3.32) to obtain a cubic equation with no quadratic term. Making this substitution into (4.4) gives us

$$x^2 = z - \frac{\lambda^2 I_\alpha}{3\,GJ}. \tag{4.5}$$

When we apply the Cardano Formulae to this z using (4.2), we will have that

$$x^2 = u - v - \frac{\lambda^2 I_\alpha}{3\,GJ}. \tag{4.6}$$

Solving for x, we find the first pair of solutions of the characteristic equation (3.27)

$$x_{1,2} = \pm\sqrt{u - v - \frac{\lambda^2 I_\alpha}{3\,GJ}}, \tag{4.7}$$

where u and v are as in (4.3) and the p and q in these expressions can be found upon comparison of (4.1) with (3.37).

We now proceed to make the necessary preliminary calculations. Upon inspection of (4.3), we find that we need to calculate $(q/2)^2$, $(p/3)^3$. For $(q/2)^2$ we have

$$\left(\frac{q}{2}\right)^2 = \left(\frac{I_\alpha}{3GJ}\right)^6 \lambda^{12} + \frac{I_\alpha^3(I_\alpha m - 3\Delta)}{3^4 EI(GJ)^4} \lambda^{10} + \left(\frac{I_\alpha m - 3\Delta}{6EI\,GJ}\right)^2 \lambda^8, \tag{4.8}$$

where q was found from (3.37). Similarly we obtain p from (3.37) and calculate

$$\left(\frac{p}{3}\right)^3 = -\left[\left(\frac{I_\alpha}{3GJ}\right)^6 \lambda^{12} + \left(\frac{I_\alpha}{3GJ}\right)^4 \left(\frac{m}{EI}\right) \lambda^{10}\right.$$

$$\left. + \frac{1}{3}\left(\frac{I_\alpha}{3GJ}\right)^2 \left(\frac{m}{EI}\right)^2 \lambda^8 + \left(\frac{m}{3EI}\right)^3 \lambda^6\right]. \tag{4.9}$$

Finally, we sum (4.8) and (4.9) and simplify as follows:

$$\left(\frac{q}{2}\right)^2 + \left(\frac{p}{3}\right)^3 = \left[\frac{I_\alpha^3(I_\alpha m - 3\Delta)}{3^4 EI(GJ)^4} - \frac{I_\alpha^4 m}{3^4 EI(GJ)^4}\right] \lambda^{10}$$

$$+ \left[\left(\frac{I_\alpha m - 3\Delta}{6EI\,GJ}\right)^2 - \frac{1}{3}\left(\frac{I_\alpha}{3GJ}\right)^2 \left(\frac{m}{EI}\right)^2\right] \lambda^8 - \left(\frac{m}{3EI}\right)^3 \lambda^6$$

$$= -\frac{I_\alpha^3 \Delta}{3^3 EI(GJ)^4} \lambda^{10} - \frac{I_\alpha^2 m^2 + 18 I_\alpha m\Delta - 27\Delta^2}{3^3 4(EI\,GJ)^2} \lambda^8 - \left(\frac{m}{3EI}\right)^3 \lambda^6. \tag{4.10}$$

Notice that the terms containing λ^{12} cancelled each other out. By simplifying the second term's numerator, we have

$$I_\alpha^2 m^2 + 18 I_\alpha m\Delta - 3^3\Delta^2 = I_\alpha^2 m^2 + 18 I_\alpha m(I_\alpha m - S^2) - 3^3(I_\alpha m - S^2)^2$$

$$= -8 I_\alpha^2 m^2 + 36 I_\alpha m S^2 - 3^3 S^4. \tag{4.11}$$

20

Using (4.11), we reduce (4.10) to the form

$$Q \equiv \left(\frac{q}{2}\right)^2 + \left(\frac{p}{3}\right)^2$$

(4.12)

$$= -\left[\frac{I_\alpha^3 \Delta}{3^3 EI \, (GJ)^4}\right] \lambda^{10} + \left[\frac{8I_\alpha^2 m^2 - 36 I_\alpha m S^2 + 27 S^4}{3^3 4 (EI \, GJ)^2}\right] \lambda^8 - \left(\frac{m}{3EI}\right)^3 \lambda^6.$$

Finally, we express u of (4.3) in terms of the parameters of the problem by comparing (4.1) and (3.37), and have

$$u = \sqrt[3]{-\left(\frac{I_\alpha}{3GJ}\right)^3 \lambda^6 - \left(\frac{I_\alpha m - 3\Delta}{6EI \, GJ}\right) \lambda^4 + \sqrt{Q}}, \qquad (4.13)$$

where Q is defined in (4.12). Similarly we express v as

$$v = \sqrt[3]{\left(\frac{I_\alpha}{3GJ}\right)^3 \lambda^6 - \left(\frac{I_\alpha m - 3\Delta}{6EI \, GJ}\right) \lambda^4 + \sqrt{Q}}, \qquad (4.14)$$

Notice that the only difference between u and v is the sign in the first term under the cubed root. Using (4.7), we can write the first pair of solutions x_1 and x_2 as

$$x_1 = \left[u - v - \left(\frac{I_\alpha}{3GJ}\right) \lambda^2\right]^{1/2}, \qquad (4.15)$$

$$x_2 = -\left[u - v - \left(\frac{I_\alpha}{3GJ}\right) \lambda^2\right]^{1/2}. \qquad (4.16)$$

To find the other two pairs of solutions, we investigate the derivation of Cardano's Formulae as described in [34]. A careful analysis shows that u and v are found using the principle cubed root. In what follows, the u–part corresponding to the first and second roots will be denoted as u_1, and the corresponding v–part will be denoted by v_1. The u–parts corresponding to the second and third roots of the cubic equation can be given by

$$u_2 = e^{i2\pi/3} u_1, \quad u_3 = e^{i4\pi/3} u_1. \qquad (4.17)$$

21

The v–parts corresponding to the second and third roots of the cubic equation can be given by

$$v_2 = e^{-i2\pi/3}v_1, \quad v_3 = e^{-i4\pi/3}v_1. \tag{4.18}$$

Substituting these results in the formulae similar to (4.7) gives us formulae for the other four solutions to the characteristic equation. We present below the formulae for the remaining four roots of the characteristic equation

$$x_3 = \left[e^{i2\pi/3}u_1 - e^{-i2\pi/3}v_1 - \left(\frac{I_\alpha}{3GJ} \right) \lambda^2 \right]^{1/2}, \tag{4.19}$$

$$x_4 = - \left[e^{i2\pi/3}u_1 - e^{-i2\pi/3}v_1 - \left(\frac{I_\alpha}{3GJ} \right) \lambda^2 \right]^{1/2}, \tag{4.20}$$

$$x_5 = \left[e^{i4\pi/3}u_1 - e^{-i4\pi/3}v_1 - \left(\frac{I_\alpha}{3GJ} \right) \lambda^2 \right]^{1/2}, \tag{4.21}$$

and

$$x_6 = - \left[e^{i4\pi/3}u_1 - e^{-i4\pi/3}v_1 - \left(\frac{I_\alpha}{3GJ} \right) \lambda^2 \right]^{1/2}. \tag{4.22}$$

Recall that the exponentials in the formulae for the roots can be written as

$$e^{i2\pi/3} = -\frac{1}{2} + i\frac{\sqrt{3}}{2} = e^{-i4\pi/3}, \quad e^{i4\pi/3} = -\frac{1}{2} - i\frac{\sqrt{3}}{2} = e^{-i2\pi/3}, \tag{4.23}$$

and these alternate expressions will be used for later calculations.

We could easily make all necessary substitutions into (4.7) and (4.19)–(4.22) to obtain exact solutions to the characteristic equation in terms of the parameters of the system. However, they are extremely complex, and are not convenient to us in such a form. We will use methods of asymptotic analysis to rewrite each root x_i in the form

$$x_i = f_i(\lambda) + c_{1i} + c_{2i}\lambda^{-n_i} + O(\lambda^{-m_i}), \quad i = 1, 2, \ldots 6, \tag{4.24}$$

where f_i is a linear combination of positive powers of λ, c_{ji} are constants, and n_i, m_i are real numbers such that $0 < n_i < m_i$.

22

Let us use the following notations:

$$v_i = \sqrt[3]{a_i\lambda^6 + c_i\lambda^4 + \sqrt{b_i\lambda^{10} + d_i\lambda^8 + e_i\lambda^6}}, \quad i = 1, 2, 3, \tag{4.25}$$

$$u_i = \sqrt[3]{-a_i\lambda^6 - c_i\lambda^4 + \sqrt{b_i\lambda^{10} + d_i\lambda^8 + e_i\lambda^6}}, \quad i = 1, 2, 3, \tag{4.26}$$

where the precise values of the constants can be found by comparison with formulae (4.25), (4.26) and (4.13),(4.14). We note that in the rest of this chapter, we will use the *binomial theorem*

$$(1 + x)^n = 1 + nx + \frac{n(n-1)}{2!}x^2 + \frac{n(n-1)(n-2)}{3!}x^3 + \dots. \tag{4.27}$$

We begin with the square root common to both (4.25) and (4.26). Without misunderstanding, we omit the subscript "i" for the rest of this section. If we apply the binomial theorem and simplify, we will have

$$\begin{aligned}
\sqrt{b\lambda^{10} + d\lambda^8 + e\lambda^6} &= \sqrt{b}\lambda^5\sqrt{1 + (d/b)\lambda^{-2} + (e/b)\lambda^{-4}} \\
&\equiv \sqrt{b}\lambda^5[1 + d'\lambda^{-2} + e'\lambda^{-4}]^{1/2} \\
&= \sqrt{b}\lambda^5[1 + (1/2)[d'\lambda^{-2} + e'\lambda^{-4}] - (1/8)[d'\lambda^{-2} + e'\lambda^{-4}]^2 + \dots \\
&= \sqrt{b}\lambda^5[1 + (1/2)d'\lambda^{-2} + O(\lambda^{-4})] \\
&= \sqrt{b}\lambda^5 + (\sqrt{b}d'/2)\lambda^3 + O(\lambda).
\end{aligned}$$
$$\tag{4.28}$$

In (4.28), we have made the substitution $d' = d/b$, and $d' = e/b$. Making further substitutions $\sqrt{b} = b'$ and $d'' = \sqrt{b}d'/2$ yields the result

$$\sqrt{b\lambda^{10} + d\lambda^8 + e\lambda^6} = b'\lambda^5 + d''\lambda^3 + O(\lambda). \tag{4.29}$$

Note, that in (4.29) there are no terms containing either λ^4 or λ^2. Substituting (4.29) into (4.25) will give an asymptotic expression for v

$$v = \sqrt[3]{a\lambda^6 + c\lambda^4 + b'\lambda^5 + d''\lambda^3 + O(\lambda)}. \tag{4.30}$$

In the calculation below, it suffices to keep the accuracy up to λ^4. We have

$$
\begin{aligned}
v &= (a\lambda^6 + b'\lambda^5 + c\lambda^4)^{1/3}(1 + O(\lambda^{-3})) \\
&= \sqrt[3]{a}\lambda^2[1 + (b'/a)\lambda^{-1} + (c/a)\lambda^{-2}]^{1/3}(1 + O(\lambda^{-3})).
\end{aligned}
\tag{4.31}
$$

Setting $b'' = b'/a$ and $c' = c/a$ and applying the binomial theorem, we obtain

$$
\begin{aligned}
v &= \sqrt[3]{a}\lambda^2[1 + b''\lambda^{-1} + c'\lambda^{-2}]^{1/3}(1 + O(\lambda^{-3})) \\
&= A\lambda^2 + B\lambda + C + O(\lambda^{-1}),
\end{aligned}
\tag{4.32}
$$

where

$$
A = \sqrt[3]{a}, \qquad B = (1/3)\sqrt[3]{a}\,b'', \qquad \text{and} \qquad C = \sqrt[3]{a}\,[(1/3)c' - (1/9)(b'')^2]. \tag{4.33}
$$

The following formulae are valid for A, B, and C in terms of the system's parameters:

$$
A = \sqrt[3]{a} = \frac{I_\alpha}{3GJ}, \qquad B = \frac{1}{3}\sqrt{\frac{-I_\alpha^3\Delta}{3^3 EI(GJ)^4}}\left(\frac{I_\alpha}{3GJ}\right)^{-2}. \tag{4.34}
$$

$$
C = \frac{I_\alpha m - 3\Delta}{2EI\,GJ}\left(\frac{3GJ}{I_\alpha}\right)^2 + \frac{I_\alpha^3\Delta}{EI(GJ)^4}\frac{(GJ)^5}{I_\alpha^5} \equiv C_1 + C_2. \tag{4.35}
$$

Similar calculations reveal that

$$
u = \tilde{A}\lambda^2 + \tilde{B}\lambda + \tilde{C} + O(\lambda^{-1}). \tag{4.36}
$$

We calculate $\tilde{A}$ and $\tilde{B}$ by replacing a and c by $-a$ and $-c$ in (4.33) to obtain

$$
\tilde{A} = -A = -\frac{I_\alpha}{3GJ}, \qquad \tilde{B} = B = \frac{\sqrt{b}}{3(-a)^{2/3}} = \frac{\sqrt{b}}{3(a)^{2/3}}. \tag{4.37}
$$

For $\tilde{C}$, we obtain

$$
\tilde{C} = -\frac{c}{3a^{2/3}} - \frac{b}{9a^2} = -\frac{I_\alpha m - 3\Delta}{2EI\,GJ}\left(\frac{GJ}{I_\alpha}\right)^2 + \frac{I_\alpha^3\Delta}{EI(GJ)^4}\frac{(GJ)^5}{I_\alpha^5} = -C_1 + C_2, \tag{4.38}
$$

where C_1 and C_2 are defined in (4.35). Now we are in a position to derive asymptotic approximations to each of the six roots.

4.2 First Pair of Roots of the Characteristic Equation

We begin with the formula for x_1 (see (4.15))

$$x_1 = \sqrt{u - v - \left(\frac{I_\alpha}{3GJ}\right)\lambda^2}. \tag{4.39}$$

Substituting expressions for u and v from (4.13) and (4.14), and recalling that v is the same with the appropriate sign changes yields

$$x_1 = \left[\sqrt[3]{-\left(\frac{I_\alpha}{3GJ}\right)^3\lambda^6 - \frac{I_\alpha m - 3\Delta}{6EI\,GJ}\lambda^4 + \sqrt{Q}} - \right.$$
$$\left. \sqrt[3]{\left(\frac{I_\alpha}{3GJ}\right)^3\lambda^6 + \frac{I_\alpha m - 3\Delta}{6EI\,GJ}\lambda^4 + \sqrt{Q}} - \frac{I_\alpha}{3GJ}\lambda^2\right]^{1/2}, \tag{4.40}$$

where Q is given in (4.12). To calculate an asymptotic approximation for x_1, we substitute asymptotic approximations (4.32) and (4.36) into formula (4.39) to obtain

$$x_1 = \sqrt{\tilde{A}\lambda^2 + \tilde{B}\lambda + \tilde{C} - A\lambda^2 - B\lambda - C - \frac{I_\alpha}{3GJ}\lambda^2 + O(\lambda^{-1})}. \tag{4.41}$$

where A, B, and C are given by formulae (4.34) and (4.35) and $\tilde{A}$, $\tilde{B}$, and $\tilde{C}$ are given in (4.37) and (4.38). The results of calculations from (4.35), (4.37) and (4.38) allow us to write

$$x_1 = \sqrt{-A\lambda^2 + B\lambda - C_1 + C_2 - A\lambda^2 - B\lambda - C_1 - C_2 - \frac{I_\alpha}{3GJ}\lambda^2 + O(\lambda^{-1})}. \tag{4.42}$$

Combining like terms yields

$$x_1 = \sqrt{\left(-2A - \frac{I_\alpha}{3GJ}\right)\lambda^2 - 2C_1\,(1 + O(\lambda^{-3}))}. \tag{4.43}$$

We continue calculations to obtain an expression for x_1 in the desired form (4.24). Making the substitution $A_1 = -2A - I_\alpha(3GJ)^{-1}$ and then factoring out the term containing λ^2 leads to the result

$$x_1 = A_1^{1/2}\lambda\left[1 - \frac{2C_1}{A_1}\lambda^{-2}\right]^{1/2}(1 + O(\lambda^{-3})). \tag{4.44}$$

25

Setting $C_2 = -2C_1/A_1$ and using the binomial theorem, we obtain that

$$x_1 = A_1^{1/2}\lambda \left[1 + \frac{1}{2}C_2\lambda^{-2} - \frac{1}{8}C_2^2\lambda^{-4} + O(\lambda^{-6}) \right] \left(1 + O(\lambda^{-3}) \right). \qquad (4.45)$$

Opening the brackets in (4.45) leads to the formula

$$x_1 = A_1^{1/2}\lambda + \frac{1}{2}A_1^{1/2}C_2\lambda^{-1} + O(\lambda^{-2}), \qquad (4.46)$$

which yields an expression for x_1 in the desired form

$$x_1 = A_2\lambda + C_3\lambda^{-1} + O(\lambda^{-2}), \qquad (4.47)$$

where we have made the substitution $A_2 = A_1^{1/2}$ and $C_3 = \frac{1}{2}A_1^{1/2}$. The asymptotic expression of the second root can be given in the form

$$x_2 = -x_1 = -A_2\lambda - C_3\lambda^{-1} + O(\lambda^{-2}). \qquad (4.48)$$

It remains to find expressions for A_2 and C_3 in terms of the parameters of the system. We calculate A_2 as

$$A_2 = A_1^{1/2} = \left(-2A - \frac{I_\alpha}{3GJ} \right)^{1/2} = \left(-2\frac{I_\alpha}{3GJ} - \frac{I_\alpha}{3GJ} \right)^{1/2} = i\sqrt{\frac{I_\alpha}{GJ}}. \qquad (4.49)$$

We also calculate C_3 as

$$C_3 = \frac{1}{2}A_1^{1/2}C_2 = -\frac{C_1}{A_1^{1/2}}. \qquad (4.50)$$

Using the substitution involving A_1 and calculation (4.35) gives

$$C_3 = \frac{-(I_\alpha m - 3\Delta)}{i(2EI\,GJ)}\left(\frac{GJ}{I_\alpha} \right)^2 \frac{1}{\left(2\frac{I_\alpha}{3GJ} + \frac{I_\alpha}{3GJ} \right)^{1/2}} = \frac{i(I_\alpha m - 3\Delta)}{2EI\,GJ\left(\frac{I_\alpha}{GJ} \right)^2 \left(\frac{I_\alpha}{GJ} \right)^{1/2}}. \qquad (4.51)$$

Finally substitution of $\Delta = I_\alpha m - S^2$ yields

$$C_3 = \frac{i(I_\alpha m - 3(I_\alpha m - S^2))}{2EI\,GJ}\left(\frac{GJ}{I_\alpha} \right)^{5/2} = \frac{i(3S^2 - 2I_\alpha m)}{2EI\,GJ}\left(\frac{GJ}{I_\alpha} \right)^{5/2}. \qquad (4.52)$$

Thus, we have obtained asymptotic expressions (4.46) and (4.48) for two roots with the constants A_2 and C_3 being given in terms of the parameters of the system in (4.49) and (4.52).

26

4.3 Second Pair of Roots of the Characteristic Equation

In this section we calculate asymptotic approximations to the roots x_3 and x_4 beginning with formulae (4.19) and (4.20). We start with the expression for x_3

$$x_3 = \sqrt{e^{i2\pi/3}u - e^{-i2\pi/3}v - \frac{I_\alpha}{3GJ}\lambda^2}. \tag{4.53}$$

Using formulae (4.23), we rewrite (4.53) in the form

$$x_3 = \sqrt{\left(-\frac{1}{2} + i\frac{\sqrt{3}}{2}\right)u + \left(\frac{1}{2} + i\frac{\sqrt{3}}{2}\right)v - \frac{I_\alpha}{3GJ}\lambda^2}. \tag{4.54}$$

We separate the real and the imaginary parts in (4.54) to obtain

$$x_3 = \sqrt{\frac{1}{2}(v - u) + i\frac{\sqrt{3}}{2}(v + u) - \frac{I_\alpha}{3GJ}\lambda^2}. \tag{4.55}$$

Using asymptotic formulae (4.32) and (4.36) for u and v, we calculate

$$
\begin{aligned}
v - u &= A\lambda^2 + B\lambda + C + O(\lambda^{-1}) - \tilde{A}\lambda^2 - \tilde{B}\lambda - \tilde{C} \\
&= (A - (-A))\lambda^2 + (B - B)\lambda + (C_1 + C_2) - (-C_1 + C_2) + O(\lambda^{-1}) \\
&= 2A\lambda^2 + 2C_1 + O(\lambda^{-1}).
\end{aligned}
\tag{4.56}
$$

To derive (4.56), we have used (4.35), (4.37), and (4.38). Similarly we calculate

$$
\begin{aligned}
v + u &= (A + \tilde{A})\lambda^2 + (B + \tilde{B})\lambda + C + \tilde{C} + O(\lambda^{-1}) \\
&= (A - A)\lambda^2 + (B + B)\lambda + (C_1 + C_2) + (-C_1 + C_2) + O(\lambda^{-1}) \\
&= 2B\lambda + 2C_2 + O(\lambda^{-1}).
\end{aligned}
\tag{4.57}
$$

Now we notice that the leading term in the expression for $(v - u)$ is quadratic with respect to λ while the leading term in the expression for $(v + u)$ is linear with respect to λ. Substituting (4.56) and (4.57) into formula (4.55) for x_3, we have

$$x_3 = \sqrt{A\lambda^2 + C_1 + O(\lambda^{-1}) + i\sqrt{3}(B\lambda + C_2 + O(\lambda^{-1})) - \frac{I_\alpha}{3GJ}\lambda^2}. \tag{4.58}$$

Combining terms having the same powers of λ and substituting the expression for A from (4.34), we obtain

$$x_3 = \sqrt{\frac{I_\alpha}{3GJ}\lambda^2 - \frac{I_\alpha}{3GJ}\lambda^2 + i\sqrt{3}B\lambda + (C_1 + i\sqrt{3}C_2) + O(\lambda^{-1})}, \qquad (4.59)$$

which shows us that the terms containing λ^2 cancel. Making the substitutions $B_{3,1} = i\sqrt{3}B$, $C_{3,3} = C_1 + i\sqrt{3}C_2$, we can represent x_3 as

$$x_3 = \sqrt{B_{3,1}\lambda + C_{3,3} + O(\lambda^{-1})}. \qquad (4.60)$$

Applying the binomial theorem to (4.60) and setting $C_{3,4} = C_{3,3}/B_{3,1}$ and $B_{3,2} = (B_{3,1})^{1/2}$, we modify the representation for x_3 and have

$$\begin{aligned}
x_3 &= [B_{3,1}\lambda + C_{3,3}]^{1/2}(1 + O(\lambda^{-2})) \\
&= B_{3,1}^{1/2}\lambda^{1/2}\left[1 + \frac{C_{3,3}}{B_{3,1}}\lambda^{-1}\right]^{1/2}(1 + O(\lambda^{-2})) \qquad (4.61) \\
&= B_{3,2}\lambda^{1/2}[1 + C_{3,4}\lambda^{-1}]^{1/2}(1 + O(\lambda^{-2})).
\end{aligned}$$

After application of the binomial theorem to the factor $[1 + C_{3,4}\lambda^{-1}]^{1/2}$, we obtain

$$x_3 = B_{3,2}\lambda^{1/2}\left[1 + \frac{1}{2}C_{3,4}\lambda^{-1} + \frac{1}{8}C_{3,4}^2\lambda^{-2} + O(\lambda^{-3})\right](1 + O(\lambda^{-2})). \qquad (4.62)$$

Making the substitution $C_{3,5} = \frac{1}{2}C_{3,4}$ and simplifying (4.62), we obtain

$$x_3 = B_{3,2}\lambda^{1/2} + B_{3,2}C_{3,5}\lambda^{-1/2} + O(\lambda^{-3/2}). \qquad (4.63)$$

Replacing $B_{3,2}C_{3,5}$ with $C_{3,6}$, we obtain the desired asymptotic approximation for x_3

$$x_3 = B_{3,2}\lambda^{1/2} + C_{3,6}\lambda^{-1/2} + O(\lambda^{-3/2}). \qquad (4.64)$$

It remains to calculate expressions for $B_{3,2}$ and $C_{3,6}$ in terms of the parameters of the system. For $B_{3,2}$ we calculate by using formulae (4.34)

$$B_{3,2} = B_{3,1}^{1/2} = [i\sqrt{3}B]^{1/2} = i\left[\frac{\sqrt{3}I_\alpha\sqrt{I_\alpha\Delta}}{3\sqrt{3}(GJ)^2\sqrt{EI}} \cdot \frac{3^2(GJ)^2}{3I_\alpha^2}\right]^{1/2} = i\left[\frac{\Delta}{I_\alpha\,EI}\right]^{1/4}, \qquad (4.65)$$

where Δ is defined in (2.4). Using (4.65), we obtain for $C_{3,6}$ the following expression:

28

$$C_{3,6} = B_{3,2}C_{3,5} = \frac{1}{2}B_{3,1}^{1/2}C_{3,4} = \frac{1}{2}B_{3,1}^{1/2}\frac{C_{3,3}}{B_{3,1}} = \frac{C_{3,3}}{2B_{3,2}} = \frac{C_1 + i\sqrt{3}C_2}{2B_{3,2}}. \tag{4.66}$$

Here we have recognized that we have calculate $B_{3,1}^{1/2}$ as $B_{3,2}$ in (4.65). We calculate C_1 and C_2 separately. First we calculate C_1 using appropriate substitutions starting with results in calculation (4.35)

$$C_1 = \left(\frac{I_\alpha m - 3\Delta}{6EI\,GJ}\right)\left(\frac{1}{3}\right)\left(\frac{3GJ}{I_\alpha}\right)^2 = \frac{(I_\alpha m - 3\Delta)(GJ)}{2\,EI\,I_\alpha^2}. \tag{4.67}$$

Similarly starting from (4.35) we have

$$C_2 = \left(\frac{I_\alpha^3 \Delta}{3^3 EI(GJ)^4}\right)\left(\frac{(3GJ)^5}{9I_\alpha^5}\right) = \frac{\Delta\,GJ}{EI\,I_\alpha^2}. \tag{4.68}$$

Having C_1 and C_2, we obtain for $C_{3,6}$

$$C_{3,6} = \frac{\dfrac{(I_\alpha m - 3\Delta)(GJ)}{2(EI)(I_\alpha)^2} + i\sqrt{3}\dfrac{\Delta(GJ)}{EI(I_\alpha)^2}}{-2i\left[\dfrac{I_\alpha m - S^2}{I_\alpha\,EI}\right]^{1/4}}. \tag{4.69}$$

Keeping in mind that $x_4 = -x_3$, we complete the approximations for the second pair of roots of Eq. (3.27).

4.4 Third Pair of Roots of the Characteristic Equation

To calculate the approximations for the two remaining roots of Eq.(3.27), we will use the same approach as in Section 4.3. Thus, we briefly outline the main steps of the derivation. We recall that

$$x_5 = \sqrt{e^{i4\pi/3}u - e^{-i4\pi/3}v - \frac{I_\alpha}{3GJ}\lambda^2}. \tag{4.70}$$

Substituting (4.23) into (4.70) and separating real and imaginary parts under the square root, we have

$$x_5 = \sqrt{\frac{1}{2}(v - u) - i\frac{\sqrt{3}}{2}(v + u) - \frac{I_\alpha}{3GJ}\lambda^2}. \tag{4.71}$$

29

Substituting asymptotic formulae (4.56) and (4.57) for $(v-u)$ and $(v+u)$ respectively and collecting like terms, we obtain

$$x_5 = \sqrt{-i\sqrt{3}B\lambda + C_1 - i\sqrt{3}C_2 + O(\lambda^{-1})}. \tag{4.72}$$

It can be easily seen that

$$x_5 = \sqrt{-i\sqrt{3}B\lambda + C_1 - i\sqrt{3}C_2}\ (1 + O(\lambda^{-2})). \tag{4.73}$$

Next we substitute $B_{5,1} = -i\sqrt{3}B$ and $C_{5,3} = C_1 - i\sqrt{3}C_2$ to have

$$x_5 = \sqrt{B_{5,1}\lambda + C_{5,3}}\ (1 + O(\lambda^{-2})) = B_{5,1}^{1/2}\lambda^{1/2}\left[1 + \frac{C_{5,3}}{B_{5,1}}\lambda^{-1}\right]^{1/2}(1 + O(\lambda^{-2})). \tag{4.74}$$

Setting $B_{5,2} = B_{5,1}^{1/2}$ and $C_{5,4} = C_{5,3}/B_{5,1}$ and then using the binomial theorem, we have

$$\begin{aligned}
x_5 &= B_{5,2}\lambda^{1/2}[1 + 1/2C_{5,4}\lambda^{-1} + O(\lambda^{-2})](1 + O(\lambda^{-2})) \\
&= B_{5,2}\lambda^{1/2} + 1/2B_{5,2}C_{5,4}\lambda^{-1/2} + O(\lambda^{-3/2}).
\end{aligned} \tag{4.75}$$

Finally, we arrive at the desired expression for x_5, i.e.,

$$x_5 = B_{5,2}\lambda^{1/2} + C_{5,5}\lambda^{-1/2} + O(\lambda^{-3/2}), \tag{4.76}$$

where $C_{5,5}$ stands for $1/2B_{5,2}C_{5,4}$.

To conclude this section, we derive formulae for the coefficients in terms of the problem's structural parameters. Using formulae (4.34), we have

$$B_{5,2} = B_{5,1}^{1/2} = [-i\sqrt{3}B]^{1/2} = \left[\frac{\sqrt{3}I_\alpha\sqrt{I_\alpha\Delta}}{3\sqrt{3}(GJ)^2\sqrt{EI}} \cdot \frac{3^2(GJ)^2}{3I_\alpha^2}\right]^{1/2} = \left[\frac{\Delta}{I_\alpha\,EI}\right]^{1/4}, \tag{4.77}$$

Using formulae (4.67) and (4.68), we calculate $C_{5,5}$

$$\begin{aligned}
C_{5,5} &= \frac{1}{2}(-i\sqrt{3}B)^{1/2}\frac{C_{5,3}}{B_{5,1}} = \frac{1}{2}\frac{-i(\sqrt{3}B)^{1/2}(C_1 - i\sqrt{3}C_2)}{-i\sqrt{3}B} \\
\\
&= \frac{1}{2}\frac{(C_1 - i\sqrt{3}C_2)}{(-i\sqrt{3}B)^{1/2}} = \frac{\dfrac{(I_\alpha m - 3\Delta)(GJ)}{2(EI)(I_\alpha)^2} - i\sqrt{3}\dfrac{\Delta(GJ)}{EI(I_\alpha)^2}}{2\left[\dfrac{I_\alpha m - S^2}{I_\alpha\,EI}\right]^{1/4}}.
\end{aligned} \tag{4.78}$$

Since $x_6 = -x_5$, we have the approximation for x_6 as well.

4.5 General Solution to the Spectral Equation for the Operator Pencil

In this section, we bring the results of the last three sections together in order to write the general solution of the differential equation $\mathcal{P}(\lambda)\phi = 0$. We recall that in Sections 4.2–4.4, we have calculated the approximations to the roots of the characteristic equation (3.27). For convenience, we reproduce those results below.

$$x_{1,2} = \pm i \sqrt{\frac{I_\alpha}{GJ}} \lambda + C_3 \lambda^{-1} + O(\lambda^{-2}), \tag{4.79}$$

$$x_{3,4} = \pm i \left[\frac{\Delta}{I_\alpha EI}\right]^{1/4} \lambda^{1/2} + C_{3,6} \lambda^{-1/2} + O(\lambda^{-3/2}), \tag{4.80}$$

$$x_{5,6} = \pm \left[\frac{\Delta}{I_\alpha EI}\right]^{1/4} \lambda^{1/2} + C_{5,6} \lambda^{-1/2} + O(\lambda^{-3/2}), \tag{4.81}$$

with C_3, $C_{3,6}$, and $C_{5,5}$ begin defined in (4.52)), (4.69) and (4.78).

The general solution to the sixth order ordinary differential equation can be represented as a linear combination of exponential–like functions. To simplify subsequent calculations, we introduce the notation

$$x_{1,2} = \pm i \Gamma(\lambda), \qquad x_{3,4} = \pm i \hat{\gamma}(\lambda), \qquad x_{5,6} = \pm \gamma(\lambda), \tag{4.82}$$

where Γ, γ, and $\hat{\gamma}$ are defined by the following formulae:

$$\gamma(\lambda) = P\lambda^{1/2} + O(\lambda^{-1/2}) = P\lambda^{1/2}(1 + O(\lambda^{-1})), \tag{4.83}$$

$$\hat{\gamma}(\lambda) = P\lambda^{1/2} + O(\lambda^{-1/2}) = P\lambda^{1/2}(1 + O(\lambda^{-1})), \tag{4.84}$$

$$\Gamma(\lambda) = Q\lambda + O(\lambda^{-1}) = Q\lambda(1 + O(\lambda^{-2})), \tag{4.85}$$

$$\text{where} \quad P = \left[\frac{\Delta}{I_\alpha EI}\right]^{1/4}, \quad Q = R^{1/2} = \left[\frac{I_\alpha}{GJ}\right]^{1/2}. \tag{4.86}$$

Using notations (4.83)–(4.86), we may write the general solution Ψ of the equation $\mathcal{P}(\lambda)\phi = 0$ in the following form:

$$\Psi(\lambda, x) = \mathcal{A}(\lambda)e^{\gamma(\lambda)(x+L)} + \mathcal{B}(\lambda)e^{i\hat{\gamma}(\lambda)(x+L)} + \mathcal{C}(\lambda)e^{i\Gamma(\lambda)(x+L)} +$$
$$\mathcal{D}(\lambda)e^{-\gamma(\lambda)(x+L)} + \mathcal{E}(\lambda)e^{-i\hat{\gamma}(\lambda)(x+L)} + \mathcal{F}(\lambda)e^{-i\Gamma(\lambda)(x+L)}, \tag{4.87}$$

with $\mathcal{A}(\cdot)$, $\mathcal{B}(\cdot)$, $\mathcal{C}(\cdot)$, $\mathcal{D}(\cdot)$, $\mathcal{E}(\cdot)$, and $\mathcal{F}(\cdot)$ being arbitrary functions of λ.

CHAPTER V
The Left–Reflection Matrix

5.1 Two–step Procedure for Applying Boundary Conditions

As stated in Section 3.2, our ultimate goal is to find such values of the complex parameter λ, for which the equation $\mathcal{L}\Psi = \lambda\Phi$ has nontrivial solutions, i.e., to find the eigenvalues and eigenvectors of the operator $\mathcal{L}$. We have shown that the aforementioned problem is equivalent to the problem of finding eigenvalues and eigenfunctions of the pencil $\mathcal{P}(\cdot)$, i.e., to the problem of finding the values of λ for which the equation $\mathcal{P}(\lambda)\phi = 0$ has nontrivial solutions. This is exactly the problem we will focus on in Chapters V–VII. Thus, we are looking for a solution of the equation

$$\mathcal{P}(\lambda)\Psi = 0, \tag{5.1}$$

which can be represented in the form (4.87). More precisely, we are looking for those $\lambda \in \mathbb{C}$, for which there exist coefficients $\mathcal{A}(\lambda)$, $\mathcal{B}(\lambda)$, $\mathcal{C}(\lambda)$, $\mathcal{D}(\lambda)$, $\mathcal{E}(\lambda)$, and $\mathcal{B}(\lambda)$, such that $\Psi(x, \lambda)$ satisfies the boundary conditions given in (3.26).

Substituting Ψ into these boundary conditions gives us a linear system of six equations in six unknowns $\mathcal{A}(\cdot)$, $\mathcal{B}(\cdot)$, $\mathcal{C}(\cdot)$, $\mathcal{D}(\cdot)$, $\mathcal{E}(\cdot)$, and $\mathcal{F}(\cdot)$. Let $\mathbb{M}$ be the 6×6 matrix of coefficients from the aforementioned system. Since our system is homogeneous, it can be written as $\mathbb{M}\mathbb{Z} = 0$, where $\mathbb{Z}^T(\lambda) = \{\mathcal{A}(\lambda), \mathcal{B}(\lambda), \mathcal{C}(\lambda), \mathcal{D}(\lambda), \mathcal{E}(\lambda), \mathcal{F}(\lambda)\}$. Thus, we have to find approximations for the solutions of the equation $det\,\mathbb{M}(\lambda) = 0$. It turns out that directly finding approximations for the roots of this determinant is an extremely difficult problem. So, we suggest an alternative approach. Namely, let us introduce two 3–component vectors

$$X(\lambda) = (\mathcal{A}(\lambda), \mathcal{B}(\lambda), \mathcal{C}(\lambda))^T, \qquad Y(\lambda) = (\mathcal{D}(\lambda), \mathcal{E}(\lambda), \mathcal{F}(\lambda))^T, \tag{5.2}$$

and first select only three boundary conditions, the conditions which have to be imposed on the solution Ψ to satisfy the boundary conditions at the left end of the beam. As a result, we obtain the relation between the vectors $X(\cdot)$ and $Y(\cdot)$, which can be written in the form

$$X(\lambda) = \mathbb{R}_l(\lambda)Y(\lambda). \tag{5.3}$$

A corresponding 3×3 matrix $\mathbb{R}_l(\cdot)$ in (5.3) we call *the left–reflection matrix.* Therefore, if the vectors $X(\cdot)$ and $Y(\cdot)$ are connected through the left–reflection matrix, the corresponding function (4.87) satisfies equation (5.1) and the left end boundary conditions. Secondly, let us select only the right–end boundary conditions. We obtain from three right end boundary conditions, that the following relation between $X(\cdot)$ and $Y(\cdot)$ holds:

$$X(\lambda) = \mathbb{R}_r(\lambda)Y(\lambda), \tag{5.4}$$

where the 3×3 matrix $\mathbb{R}_r(\cdot)$ we call *the right–reflection matrix.* So, if the vectors $X(\cdot)$ and $Y(\cdot)$ are connected through relation (5.4), the corresponding function (4.87) satisfies equation (5.1) and three boundary conditions at the right end.

It can be easily verified that to satisfy all six boundary conditions, the following equation must be valid:

$$\begin{pmatrix} \mathcal{A}(\lambda) \\ \mathcal{B}(\lambda) \\ \mathcal{C}(\lambda) \\ \mathcal{D}(\lambda) \\ \mathcal{E}(\lambda) \\ \mathcal{F}(\lambda) \end{pmatrix} = \left[\begin{array}{c|c} 0 & \mathbb{R}_r(\lambda) \\ \hline \mathbb{R}_l^{-1}(\lambda) & 0 \end{array} \right] \begin{pmatrix} \mathcal{A}(\lambda) \\ \mathcal{B}(\lambda) \\ \mathcal{C}(\lambda) \\ \mathcal{D}(\lambda) \\ \mathcal{E}(\lambda) \\ \mathcal{F}(\lambda) \end{pmatrix}. \tag{5.5}$$

Eq.(5.5) is certainly equivalent to the following one:

$$\left(\mathbb{I} - \left[\begin{array}{c|c} 0 & \mathbb{R}_r \\ \hline \mathbb{R}_l^{-1} & 0 \end{array} \right] \right) \begin{pmatrix} \mathcal{A}(\lambda) \\ \mathcal{B}(\lambda) \\ \mathcal{C}(\lambda) \\ \mathcal{D}(\lambda) \\ \mathcal{E}(\lambda) \\ \mathcal{F}(\lambda) \end{pmatrix} = 0, \tag{5.6}$$

where $\mathbb{I}$ is the identity matrix. We notice that a solution of Eq.(5.6) is nontrivial if and only if

$$\det \left(\mathbb{I} - \left[\begin{array}{c|c} 0 & \mathbb{R}_r(\lambda) \\ \hline \mathbb{R}_l^{-1}(\lambda) & 0, \end{array} \right] \right) = 0, \tag{5.7}$$

or equivalently

$$\det(\mathbb{I} - \mathbb{R}_l^{-1}(\lambda)\mathbb{R}_r(\lambda)) = 0. \tag{5.8}$$

We may factor out $\mathbb{R}_l^{-1}$ and obtain

$$\det(\mathbb{R}_l^{-1}(\lambda)) \det(\mathbb{R}_l(\lambda) - \mathbb{R}_r(\lambda)) = 0, \tag{5.9}$$

so that since $\mathbb{R}_l^{-1}$ exists (as will be shown later), we obtain

$$\det(\mathbb{R}_l(\lambda) - \mathbb{R}_r(\lambda)) = 0. \tag{5.10}$$

Thus we have reduced the problem involving a 6×6 matrix to a similar problem for a 3×3 matrix. From now on, we will carry out the following steps:

• calculate the left and right–reflection matrices;

• find approximations for the roots of Eq.(5.10).

5.2   Left–Reflection Matrix

In this section, we will derive an asymptotic approximation for the left–reflection matrix $\mathbb{R}_l$. Let us substitute the general solution $\Psi(\cdot)$ given in (4.87) into each of the three left–hand boundary conditions of the operator pencil. Substituting into the first one gives us the first equation for unknown coefficients

$$\mathcal{A}(\lambda) + \mathcal{B}(\lambda) + \mathcal{C}(\lambda) + \mathcal{D}(\lambda) + \mathcal{E}(\lambda) + \mathcal{F}(\lambda) = 0. \tag{5.11}$$

Substituting $\Psi(\cdot)$ from (4.87) into the second one gives us the second equation

$$\mathcal{A}(\lambda)\gamma(\lambda) + \mathcal{B}(\lambda)i\hat{\gamma}(\lambda) + \mathcal{C}(\lambda)i\Gamma(\lambda) - \mathcal{D}(\lambda)\gamma(\lambda) - \mathcal{E}(\lambda)i\hat{\gamma}(\lambda) - \mathcal{F}(\lambda)i\Gamma(\lambda) = 0. \tag{5.12}$$

with $\gamma(\cdot)$, $\hat{\gamma}(\cdot)$, and $\Gamma(\cdot)$ being defined in (4.83)–(4.85). And lastly substituting $\Psi(\cdot)$ into the third boundary condition (3.20) yields

$$[\mathcal{A}(\lambda)\gamma^4(\lambda) + \mathcal{B}(\lambda)\hat{\gamma}^4(\lambda) + \mathcal{C}(\lambda)\Gamma^4(\lambda) + \mathcal{D}(\lambda)\gamma^4(\lambda) + \mathcal{E}(\lambda)\hat{\gamma}^4(\lambda) + \mathcal{F}(\lambda)\Gamma^4(\lambda)] + 0 = 0. \tag{5.13}$$

Rearranging the above three equations so that the functions $\mathcal{A}(\cdot)$, $\mathcal{B}(\cdot)$, and $\mathcal{C}(\cdot)$ are on one side while the functions $\mathcal{D}(\cdot)$, $\mathcal{E}(\cdot)$, and $\mathcal{F}(\cdot)$ are on the other side results in the following linear system of three equations:

$$\mathcal{A}(\lambda) + \mathcal{B}(\lambda) + \mathcal{C}(\lambda) = -\mathcal{D}(\lambda) - \mathcal{E}(\lambda) - \mathcal{F}(\lambda), \tag{5.14}$$

$$\gamma(\lambda)\mathcal{A}(\lambda) + i\hat{\gamma}(\lambda)\mathcal{B}(\lambda) + i\Gamma(\lambda)\mathcal{C}(\lambda) = \gamma(\lambda)\mathcal{D}(\lambda) + i\hat{\gamma}(\lambda)\mathcal{E}(\lambda) + i\Gamma(\lambda)\mathcal{F}(\lambda), \tag{5.15}$$

$$\gamma^4(\lambda)\mathcal{A}(\lambda) + \hat{\gamma}^4(\lambda)\mathcal{B}(\lambda) + \Gamma^4(\lambda)\mathcal{C}(\lambda) = -\gamma^4(\lambda)\mathcal{D}(\lambda) - \hat{\gamma}^4(\lambda)\mathcal{E}(\lambda) - \Gamma^4(\lambda)\mathcal{F}(\lambda). \tag{5.16}$$

Clearly, the three equation (5.14)–(5.16) can be written as one matrix equation

$$\begin{bmatrix} 1 & 1 & 1 \\ \gamma(\lambda) & i\hat{\gamma}(\lambda) & i\Gamma(\lambda) \\ \gamma^4(\lambda) & \hat{\gamma}^4(\lambda) & \Gamma^4(\lambda) \end{bmatrix} \begin{bmatrix} \mathcal{A}(\lambda) \\ \mathcal{B}(\lambda) \\ \mathcal{C}(\lambda) \end{bmatrix} = \begin{bmatrix} -1 & -1 & -1 \\ \gamma(\lambda) & i\hat{\gamma}(\lambda) & i\Gamma(\lambda) \\ -\gamma^4(\lambda) & -\hat{\gamma}^4(\lambda) & -\Gamma^4(\lambda) \end{bmatrix} \begin{bmatrix} \mathcal{D}(\lambda) \\ \mathcal{E}(\lambda) \\ \mathcal{F}(\lambda) \end{bmatrix}. \tag{5.17}$$

Thus we have a matrix equation in the form

$$\mathbb{A}(\lambda)X(\lambda) = \mathbb{B}(\lambda)Y(\lambda), \tag{5.18}$$

35

where the vectors $X(\cdot)$ and $Y(\cdot)$ are defined in (5.2). If we solve Eq.(5.18) for $X(\cdot)$

$$X(\lambda) = \mathbb{A}^{-1}(\lambda)\mathbb{B}(\lambda)Y(\lambda), \tag{5.19}$$

then we observe that the left–reflection matrix as described in (5.3) can be given as

$$\mathbb{R}_l(\lambda) = \mathbb{A}^{-1}(\lambda)\mathbb{B}(\lambda). \tag{5.20}$$

While the straightforward calculation of the above left–reflection matrix is possible, we exploit the similarity of the entries of $\mathbb{A}$ and $\mathbb{B}$ to make the calculation easier. We notice that

$$\mathbb{B}(\lambda) = -\mathbb{A}(\lambda) + \mathbb{V}(\lambda), \tag{5.21}$$

where

$$\mathbb{V}(\lambda) = 2 \begin{bmatrix} 0 & 0 & 0 \\ \gamma(\lambda) & i\hat{\gamma}(\lambda) & i\Gamma(\lambda) \\ 0 & 0 & 0 \end{bmatrix}. \tag{5.22}$$

Thus the calculation of $\mathbb{R}_l(\cdot)$ can be simplified, for using this expression for $\mathbb{B}(\cdot)$ in terms of $\mathbb{A}(\cdot)$ we have that

$$\mathbb{R}_l(\lambda) = \mathbb{A}^{-1}(\lambda)\mathbb{B}(\lambda) = \mathbb{A}^{-1}(\lambda)(-\mathbb{A}(\lambda) + \mathbb{V}(\lambda)) = -\mathbb{I} + \mathbb{A}^{-1}(\lambda)\mathbb{V}(\lambda). \tag{5.23}$$

This alternate expression for $\mathbb{R}_l(\cdot)$ will make its calculation easier because only the middle column of $\mathbb{A}^{-1}(\cdot)$ is needed for the calculation of $\mathbb{A}^{-1}(\cdot)\mathbb{V}(\cdot)$ since $\mathbb{V}(\cdot)$ has only one non–zero row. So recalling that

$$\mathbb{A}(\lambda) = \begin{bmatrix} 1 & 1 & 1 \\ \gamma(\lambda) & i\hat{\gamma}(\lambda) & i\Gamma(\lambda) \\ \gamma^4(\lambda) & \hat{\gamma}^4(\lambda) & \Gamma^4(\lambda) \end{bmatrix}, \tag{5.24}$$

36

we calculate asymptotic approximations to the middle column of $\mathbb{A}^{-1}(\cdot)$ using Cramer's Rule. First we calculate $det\, \mathbb{A}$ by expansion with respect to the bottom row entries

$$(\det \mathbb{A})(\lambda) = \gamma^4(\lambda) \begin{vmatrix} 1 & 1 \\ i\hat{\gamma}(\lambda) & i\Gamma(\lambda) \end{vmatrix} - \hat{\gamma}^4(\lambda) \begin{vmatrix} 1 & 1 \\ i\gamma(\lambda) & i\Gamma(\lambda) \end{vmatrix} + \Gamma^4(\lambda) \begin{vmatrix} 1 & 1 \\ \gamma(\lambda) & i\hat{\gamma}(\lambda) \end{vmatrix}.$$
(5.25)

Substituting expressions (4.83)–(4.85) into (5.25), we find that the terms containing γ^4 and $\hat{\gamma}^4$ behave as $O(\lambda^3)$ when $|\lambda| \to \infty$, while the term containing Γ^4 behaves as $O(\lambda^{4.5})$ when $|\lambda| \to \infty$. The latter fact means that we can proceed as follows:

$$(\det \mathbb{A})(\lambda) = \Gamma^4(\lambda)(i\hat{\gamma}(\lambda) - \gamma(\lambda)) + O(\lambda^3) = \Gamma^4(\lambda)(i\hat{\gamma}(\lambda) - \gamma(\lambda))(1 + O(\lambda^{-1.5})).$$ (5.26)

Substituting formulae (4.83) and (4,84) into (5.26) leads to

$$\begin{aligned}(\det \mathbb{A})(\lambda) &= \Gamma^4(\lambda)[P\lambda^{1/2}(i - 1) + O(\lambda^{-1/2})](1 + O(\lambda^{-1.5})) \\ &= \Gamma^4(\lambda)[P\lambda^{1/2}(i - 1)](1 + O(\lambda^{-1})).\end{aligned}$$
(5.27)

Now we proceed to find each specific entry of the middle column of the matrix $\mathbb{A}^{-1}(\cdot)$ given by (5.24). Let C_{2j}, $j = 1, 2, 3$, be a cofactor corresponding to the second row, and the j–th entry of $\mathbb{A}(\cdot)$. Beginning with the first entry of $\mathbb{A}^{-1}(\cdot)$, we have

$$\mathbb{A}_{12}^{-1}(\lambda) = \frac{C_{21}}{(\det \mathbb{A})(\lambda)} = \frac{-1}{(\det \mathbb{A})(\lambda)} \begin{vmatrix} 1 & 1 \\ \hat{\gamma}^4(\lambda) & \Gamma^4(\lambda) \end{vmatrix} = \frac{-(\Gamma^4(\lambda) - \hat{\gamma}^4(\lambda))}{\Gamma^4(\lambda)[P\lambda^{1/2}(i - 1)](1 + O(\lambda^{-1/2}))}.$$
(5.28)

Recalling that $\Gamma(\cdot)$ behaves as $O(\lambda)$ while $\hat{\gamma}(\cdot)$ behaves as $O(\lambda^{1/2})$, we may rewrite the numerator so that

$$\mathbb{A}_{12}^{-1}(\lambda) = \frac{-\Gamma^4(\lambda)(1 + O(\lambda^{-2}))}{\Gamma^4(\lambda)[P\lambda^{1/2}(i - 1) + O(\lambda^{-1/2})]} = \frac{1 + O(\lambda^{-2})}{(1 - i)P\lambda^{1/2}(1 + O(\lambda^{-1}))}.$$ (5.29)

Now using the fact that

$$\frac{1}{1 + O(\lambda^m)} = 1 + O(\lambda^m) + O(\lambda^{2m}) + \ldots = 1 + O(\lambda^m), \; m < 0,$$ (5.30)

we can finally write

$$\mathbb{A}_{12}^{-1}(\lambda) = \frac{1}{(1 - i)} P^{-1}\lambda^{-1/2}(1 + O(\lambda^{-1})).$$ (5.31)

Thus we have obtained an asymptotic expression of $\mathbb{A}_{12}^{-1}$. In what follows, it is convenient to use the notation

$$\hat{\omega}_{ij}(\lambda) = 1 + O(\lambda^{-1}), \tag{5.32}$$

which means that on the intersection of the $i-th$ row and the $j-th$ column, there is a factor $1 + O(\lambda^{-1})$. Thus we may finally write that

$$\mathbb{A}_{12}^{-1} = \frac{1+i}{2} P^{-1} \lambda^{-1/2} \hat{\omega}_{12}. \tag{5.33}$$

Calculations of the other two needed entries of $\mathbb{A}^{-1}(\cdot)$ will proceed in a similar manner. Calculating $\mathbb{A}_{22}^{-1}(\cdot)$, we have

$$
\begin{aligned}
\mathbb{A}_{22}^{-1}(\lambda) &= \frac{C_{22}}{(\det \mathbb{A})(\lambda)} = \frac{(-1)^4}{\Gamma^4(\lambda)[P\lambda^{1/2}(i-1)]\hat{\omega}_{22}(\lambda)} \begin{vmatrix} 1 & 1 \\ \gamma^4(\lambda) & \Gamma^4(\lambda) \end{vmatrix} \\
&= \frac{\Gamma^4(\lambda) - \gamma^4(\lambda)}{\Gamma^4(\lambda)[P\lambda^{1/2}(i-1)]\hat{\omega}_{22}(\lambda)}.
\end{aligned}
\tag{5.34}
$$

Substituting expressions (4.83) into the numerator we find that $\gamma^4(\cdot)$ behaves as $O(\lambda^2)$. Thus we may rewrite the numerator and have

$$\mathbb{A}_{22}^{-1}(\lambda) = -\frac{i+1}{2} P^{-1} \lambda^{-1/2} \hat{\omega}_{22}(\lambda). \tag{5.35}$$

Calculating the remaining entry, we have

$$
\begin{aligned}
\mathbb{A}_{32}^{-1}(\lambda) &= \frac{C_{23}}{(\det \mathbb{A})(\lambda)} = \frac{-1}{\Gamma^4(\lambda)[P\lambda^{1/2}(i-1)](1+O(\lambda^{-1}))} \begin{vmatrix} 1 & 1 \\ \gamma^4(\lambda) & \hat{\gamma}^4(\lambda) \end{vmatrix} \\
&= \frac{\gamma^4(\lambda) - \hat{\gamma}^4(\lambda)}{\Gamma^4(\lambda)[P\lambda^{1/2}(i-1)](1+O(\lambda^{-1}))}.
\end{aligned}
\tag{5.36}
$$

Using formulae (4.83) and (4.84), we can see that $\gamma^4(\lambda) - \hat{\gamma}^4(\lambda) = O(\lambda)$. Substituting the expression for $\Gamma(\cdot)$ from (4.85) and simplifying, we obtain

$$\mathbb{A}_{32}^{-1}(\lambda) = \frac{O(\lambda)}{\tilde{P}\lambda^4(1+O(\lambda^{-2}))[P\lambda^{1/2}(i-1)](1+O(\lambda^{-1}))} = O(\lambda^{-3.5}). \tag{5.37}$$

Using formula (5.22) for $\mathbb{V}$ and (5.23) for $\mathbb{R}_l$, we obtain

$$\mathbb{R}_l(\lambda) = -\mathbb{I} + \mathbb{A}^{-1}(\lambda)\mathbb{V}(\lambda)$$

$$= \begin{bmatrix} -1 & 0 & 0 \\ 0 & -1 & 0 \\ 0 & 0 & -1 \end{bmatrix} + 2 \begin{bmatrix} * & \mathbb{A}_{12}^{-1}(\lambda) & * \\ * & \mathbb{A}_{22}^{-1}(\lambda) & * \\ * & \mathbb{A}_{32}^{-1}(\lambda) & * \end{bmatrix} \begin{bmatrix} 0 & 0 & 0 \\ \gamma(\lambda) & i\hat{\gamma}(\lambda) & i\Gamma(\lambda) \\ 0 & 0 & 0 \end{bmatrix} \tag{5.38}$$

$$= \begin{bmatrix} -1 & 0 & 0 \\ 0 & -1 & 0 \\ 0 & 0 & -1 \end{bmatrix} + 2 \begin{bmatrix} \gamma(\lambda)\mathbb{A}_{12}^{-1}(\lambda) & i\hat{\gamma}(\lambda)\mathbb{A}_{12}^{-1}(\lambda) & i\Gamma(\lambda)\mathbb{A}_{12}^{-1}(\lambda) \\ \gamma(\lambda)\mathbb{A}_{22}^{-1}(\lambda) & i\hat{\gamma}(\lambda)\mathbb{A}_{22}^{-1}(\lambda) & i\Gamma(\lambda)\mathbb{A}_{22}^{-1}(\lambda) \\ \gamma(\lambda)\mathbb{A}_{32}^{-1}(\lambda) & i\hat{\gamma}(\lambda)\mathbb{A}_{32}^{-1}(\lambda) & i\Gamma(\lambda)\mathbb{A}_{32}^{-1}(\lambda) \end{bmatrix}.$$

In (5.38), we have used the notation "*" for those entries, which are immaterial for us. Finally, we have the following representation for the left–reflection matrix:

$$\mathbb{R}_l(\lambda) = \begin{bmatrix} -1 + 2\gamma(\lambda)\mathbb{A}_{12}^{-1}(\lambda) & 2i\hat{\gamma}(\lambda)\mathbb{A}_{12}^{-1}(\lambda) & 2i\Gamma(\lambda)\mathbb{A}_{12}^{-1}(\lambda) \\ 2\gamma(\lambda)\mathbb{A}_{22}^{-1}(\lambda) & -1 + 2i\hat{\gamma}(\lambda)\mathbb{A}_{22}^{-1}(\lambda) & 2i\Gamma(\lambda)\mathbb{A}_{22}^{-1}(\lambda) \\ 2\gamma(\lambda)\mathbb{A}_{32}^{-1}(\lambda) & 2i\hat{\gamma}(\lambda)\mathbb{A}_{32}^{-1}(\lambda) & -1 + 2i\Gamma(\lambda)\mathbb{A}_{32}^{-1}(\lambda) \end{bmatrix}. \tag{5.39}$$

Using formulae (4.83) and (5.33) for $\gamma(\cdot)$ and $\mathbb{A}_{12}^{-1}(\cdot)$, we can write

$$\begin{aligned} \gamma(\lambda)\mathbb{A}_{12}^{-1}(\lambda) &= P\lambda^{1/2}(1 + O(\lambda^{-1}))\frac{(1 + O(\lambda^{-2}))}{(1-i)P\lambda^{1/2}(1 + O(\lambda^{-1}))} \\ &= \frac{1+i}{2}\,(1 + O(\lambda^{-1})). \end{aligned} \tag{5.40}$$

A similar calculation for $\hat{\gamma}(\cdot)\mathbb{A}_{12}^{-1}(\cdot)$ yields

$$
\begin{aligned}
\hat{\gamma}(\lambda)\mathbb{A}_{12}^{-1}(\lambda) =& P\lambda^{1/2}(1+O(\lambda^{-1/2}))\frac{(1+O(\lambda^{-2}))}{(1-i)P\lambda^{1/2}(1+O(\lambda^{-1}))} \\
=& \frac{1+i}{2}\ (1+O(\lambda^{-1})).
\end{aligned}
\tag{5.41}
$$

Using (4.85) and (5.33), we calculate $\Gamma(\cdot)\mathbb{A}_{12}^{-1}(\cdot)$ and have

$$
\begin{aligned}
\Gamma(\lambda)\mathbb{A}_{12}^{-1}(\lambda) =& \sqrt{I_\alpha/GJ}\lambda(1+O(\lambda^{-2}))\frac{(1+O(\lambda^{-2}))}{(1-i)P\lambda^{1/2}(1+O(\lambda^{-1}))} \\
=& \sqrt{\frac{I_\alpha}{GJ}}P^{-1}\lambda^{1/2}\frac{(i+1)}{2}\ (1+O(\lambda^{-1})).
\end{aligned}
\tag{5.42}
$$

Now we move on to the second row of the matrix $\mathbb{R}_l$. Using (4.84) and (5.35), we calculate that

$$
\hat{\gamma}(\lambda)\mathbb{A}_{22}^{-1}(\lambda) = -P\lambda^{1/2}\frac{(i+1)}{2}P^{-1}\lambda^{-1/2}\hat{\omega}_{22} = -\frac{(i+1)\hat{\omega}_{22}}{2}.
\tag{5.43}
$$

Thus we substitute this expression into the entire expression for $(\mathbb{R}_l)_{22}$ in (5.39) and simplify to have

$$
(\mathbb{R}_l)_{22}(\lambda) = -1 + 2\mathbb{A}_{22}^{-1}(\lambda)i\hat{\gamma}(\lambda) = -1 - i(i+1)\hat{\omega}_{22}(\lambda) = -i\hat{\omega}_{22}(\lambda).
\tag{5.44}
$$

Now considering the entry $(\mathbb{R}_l)_{21}$, we substitute (4.84) and (5.35) and have

$$
(\mathbb{R}_l)_{21}(\lambda) = 2\gamma(\lambda)\mathbb{A}_{22}^{-1}(\lambda) = -[P\lambda^{1/2}\hat{\omega}_{22}(\lambda)](i+1)P^{-1}\lambda^{-1/2}\hat{\omega}_{22}(\lambda) = -(i+1)\hat{\omega}_{22}(\lambda).
\tag{5.45}
$$

Turning now to $(\mathbb{R}_l)_{23}(\cdot)$, we calculate a part of it by substitution (4.85) and (5.35)

$$
\begin{aligned}
\Gamma(\lambda)\mathbb{A}_{22}^{-1}(\lambda) =& \left[\sqrt{\frac{I_\alpha}{GJ}}\lambda(1+O(\lambda^{-2}))\right]\left[\frac{-(i+1)}{2}P^{-1}\lambda^{-1/2}\hat{\omega}_{22}(\lambda)\right] \\
=& -\frac{(i+1)}{2}\sqrt{\frac{I_\alpha}{GJ}}P^{-1}\lambda^{1/2}\hat{\omega}_{22}(\lambda).
\end{aligned}
\tag{5.46}
$$

Thus we may calculate the entry $(\mathbb{R}_l)_{23}(\cdot)$ as

$$
(\mathbb{R}_l)_{23}(\lambda) = 2i\Gamma(\lambda)\mathbb{A}_{22}^{-1}(\lambda) = (1-i)\sqrt{\frac{I_\alpha}{GJ}}P^{-1}\lambda^{1/2}\hat{\omega}_{23}(\lambda).
\tag{5.47}
$$

40

Finally we evaluate the third row of $\mathbb{R}_l(\cdot)$. We calculate a portion of the first entry using (4.83) as well as (5.36) to find that

$$\gamma(\lambda)\mathbb{A}_{32}^{-1}(\lambda) = P\lambda^{1/2}(1 + O(\lambda^{-1}))O(\lambda^{-3.5}) = O(\lambda^{-3}). \tag{5.48}$$

Similarly, we calculate a portion of the second entry as

$$\hat{\gamma}(\lambda)\mathbb{A}_{32}^{-1}(\lambda) = P\lambda^{1/2}(1 + O(\lambda^{-1}))O(\lambda^{-3.5}) = O(\lambda^{-3}). \tag{5.49}$$

And finally we calculate a portion of the last entry as

$$\Gamma(\lambda)\mathbb{A}_{32}^{-1}(\lambda) = \sqrt{\frac{I_\alpha}{GJ}}\lambda(1 + O(\lambda^{-2}))O(\lambda^{-3.5}) = O(\lambda^{-2.5}). \tag{5.50}$$

Substituting the results of (5.40)–(5.42), (5.44)–(5.46), and (5.48)–(5.50) into (5.39) for $\mathbb{R}_l(\cdot)$ and simplifying, we obtain the following asymptotic approximation to the left–reflection matrix:

$$\mathbb{R}_l(\lambda) = \begin{bmatrix} i\hat{\omega}_{11}(\lambda) & (i-1)\hat{\omega}_{12}(\lambda) & \sqrt{\dfrac{I_\alpha}{GJ}}P^{-1}(i-1)\lambda^{1/2}\hat{\omega}_{13}(\lambda) \\[2mm] -(i+1)\hat{\omega}_{21}(\lambda) & -i\hat{\omega}_{22}(\lambda) & \sqrt{\dfrac{I_\alpha}{GJ}}P^{-1}(1-i)\lambda^{1/2}\hat{\omega}_{23}(\lambda) \\[2mm] O(\lambda^{-3}) & O(\lambda^{-3}) & -1 + O(\lambda^{-2.5}) \end{bmatrix}. \tag{5.51}$$

CHAPTER VI

Right–Reflection Matrix

Now we will look for the right–reflection matrix $\mathbb{R}_r$ by substituting the general solution $\Psi(\cdot)$ from (4.87) into the right end boundary conditions (3.21) and (3.24). In what follows, it is convenient to introduce new notation

$$
\begin{aligned}
\exp\{\gamma(\lambda)L\} &\equiv e(\lambda) \equiv e^{\gamma(\lambda)L}, \\
\exp\{i\hat{\gamma}(\lambda)L\} &\equiv \hat{e}(\lambda) \equiv e^{i\hat{\gamma}(\lambda)L}, \\
\exp\{i\Gamma(\lambda)L\} &\equiv e_+(\lambda) \equiv e^{i\Gamma(\lambda)L}.
\end{aligned}
\tag{6.1}
$$

We also recall that by (4.86)

$$
P = \left[\frac{\Delta}{I_\alpha\,EI}\right]^{1/4}, \quad R^{1/2} = Q = \left[\frac{I_\alpha}{GJ}\right]^{1/2}.
\tag{6.2}
$$

Beginning with the first right end boundary condition $\Psi'''(0) = 0$ (as stated in (2.15)), we substitute $\Psi(\cdot)$ from (4.87) and use definition (6.1) to obtain

$$
\begin{aligned}
&\gamma^3(\lambda)\mathcal{A}(\lambda)e(\lambda) - i\hat{\gamma}^3(\lambda)\mathcal{B}(\lambda)\hat{e}(\lambda) - i\Gamma^3(\lambda)\mathcal{C}(\lambda)e_+(\lambda) - \\
&\gamma^3(\lambda)\mathcal{D}(\lambda)e(\lambda)^{-1} + i\hat{\gamma}^3(\lambda)\mathcal{E}(\lambda)\hat{e}(\lambda)^{-1} + i\Gamma^3(\lambda)\mathcal{F}(\lambda)e_+(\lambda)^{-1} = 0.
\end{aligned}
\tag{6.3}
$$

To proceed, we need the following estimates:

$$
\frac{\hat{\gamma}(\lambda)}{\gamma(\lambda)} = (1 + O(\lambda^{-1})), \quad \frac{\Gamma(\lambda)}{\gamma(\lambda)} = \frac{Q}{P}\lambda^{1/2}\,(1 + O(\lambda^{-1})).
\tag{6.4}
$$

Thus after dividing (6.3) by $\gamma^3(\lambda)$ and substituting (6.4) into the result, we obtain

$$
\mathcal{A}(\lambda)e(\lambda) - i(1 + O(\lambda^{-1}))\mathcal{B}(\lambda)\hat{e}(\lambda) - i\frac{Q^3}{P^3}\lambda^{3/2}(1 + O(\lambda^{-1}))\mathcal{C}(\lambda)e_+(\lambda)
$$

$$
-\mathcal{D}(\lambda)e(\lambda)^{-1} + i(1 + O(\lambda^{-1}))\mathcal{E}(\lambda)\hat{e}(\lambda)^{-1} + i\frac{Q^3}{P^3}\lambda^{3/2}(1 + O(\lambda^{-1}))\mathcal{F}(\lambda)e_+(\lambda)^{-1} = 0.
$$

$$
\tag{6.5}
$$

Let us leave the terms containing $\mathcal{A}(\cdot)$, $\mathcal{B}(\cdot)$ and $\mathcal{C}(\cdot)$ on the left side of equation (6.5) while moving the terms with $\mathcal{D}(\cdot)$, $\mathcal{E}(\cdot)$, and $\mathcal{F}(\cdot)$ to the right side to obtain

$$
\mathcal{A}(\lambda)e(\lambda) - i(1 + O(\lambda^{-1}))\mathcal{B}(\lambda)\hat{e}(\lambda) - i\frac{Q^3}{P^3}\lambda^{3/2}(1 + O(\lambda^{-1}))\mathcal{C}(\lambda)e_+(\lambda) =
$$

$$
\mathcal{D}(\lambda)e(\lambda)^{-1} - i(1 + O(\lambda^{-1}))\mathcal{E}(\lambda)\hat{e}(\lambda)^{-1} - i\frac{Q^3}{P^3}\lambda^{3/2}(1 + O(\lambda^{-1}))\mathcal{F}(\lambda)e_+(\lambda)^{-1}.
$$

$$
\tag{6.6}
$$

Now we turn to the second right–hand side boundary condition of the operator pencil given in (3.21) as

$$EI\varphi_0''(0) + g_h i\lambda\varphi_0'(0) = 0. \tag{6.7}$$

Substituting in $\Psi(\cdot)$ from (4.87) and using notation (6.1), we obtain

$$EI[\gamma^2(\lambda)\mathcal{A}(\lambda)e(\lambda) - \hat{\gamma}^2(\lambda)\mathcal{B}(\lambda)\hat{e}(\lambda) - \Gamma^2(\lambda)\mathcal{C}(\lambda)e_+(\lambda) + \gamma^2(\lambda)\mathcal{D}(\lambda)e(\lambda)^{-1} -$$

$$\hat{\gamma}^2(\lambda)\mathcal{E}(\lambda)\hat{e}(\lambda)^{-1} - \Gamma^2(\lambda)\mathcal{F}(\lambda)e_+(\lambda)^{-1}] + g_h i\lambda[\gamma(\lambda)\mathcal{A}(\lambda)e(\lambda) + i\hat{\gamma}(\lambda)\mathcal{B}(\lambda)\hat{e}(\lambda) +$$

$$i\Gamma(\lambda)\mathcal{C}(\lambda)e_+(\lambda) - \gamma(\lambda)\mathcal{D}(\lambda)e(\lambda)^{-1} - i\hat{\gamma}(\lambda)\mathcal{E}(\lambda)\hat{e}(\lambda)^{-1} - i\Gamma(\lambda)\mathcal{F}(\lambda)e_+(\lambda)^{-1}] = 0. \tag{6.8}$$

For the next step, we need the following approximations:

$$\frac{1}{\gamma(\lambda)} = P^{-1}\lambda^{-1/2}(1 + O(\lambda^{-1})), \qquad \frac{\hat{\gamma}(\lambda)}{\gamma^2(\lambda)} = P^{-1}\lambda^{-1/2}(1 + O(\lambda^{-1})),$$

$$\frac{\Gamma(\lambda)}{\gamma^2(\lambda)} = QP^{-2}(1 + O(\lambda^{-1})). \tag{6.9}$$

We divide (6.8) by $\gamma^2(\cdot)$ using approximations (6.4) and (6.9), and then collect together terms involving each of $\mathcal{A}(\cdot)$, $\mathcal{B}(\cdot)$, $\mathcal{C}(\cdot)$, $\mathcal{D}(\cdot)$, $\mathcal{E}(\cdot)$ and $\mathcal{F}(\cdot)$ respectively. Let $Q1$ be the coefficient for $\mathcal{A}(\cdot)$. $Q1$ has the following asymptotic approximation:

$$Q1 \equiv [EI + g_h i\lambda P^{-1}\lambda^{-1/2}(1 + O(\lambda^{-1}))]e(\lambda) = [EI + g_h P^{-1}i\lambda^{1/2}](1 + O(\lambda^{-1}))e(\lambda). \tag{6.10}$$

The coefficient for $\mathcal{B}(\cdot)$, denoted by $Q2$, has the following asymptotic approximation:

$$Q2 \equiv -\left[EI\left(\frac{\hat{\gamma}(\lambda)}{\gamma(\lambda)}\right)^2 + g_h\lambda\frac{\hat{\gamma}(\lambda)}{\gamma^2(\lambda)}\right]\hat{e}(\lambda) = [-EI - g_h P^{-1}\lambda^{1/2}](1 + O(\lambda^{-1}))\hat{e}(\lambda). \tag{6.11}$$

The coefficient $Q3$ before $\mathcal{C}(\cdot)$ can be approximated as

$$Q3 \equiv -\left[EI\left(\frac{\Gamma(\lambda)}{\gamma(\lambda)}\right)^2 + g_h\lambda\frac{\Gamma(\lambda)}{\gamma^2(\lambda)}\right]e_+(\lambda)$$

$$= [-EIQ^2P^{-2}\lambda - g_h\lambda QP^{-2}](1 + O(\lambda^{-1}))e_+(\lambda). \tag{6.12}$$

The coefficient $Q4$ before $\mathcal{D}(\cdot)$ can be approximated as

$$Q4 \equiv \left[EI - g_h i\lambda\frac{\gamma(\lambda)}{\gamma^2(\lambda)}\right]e(\lambda)^{-1} = [EI - g_h P^{-1}i\lambda^{1/2}](1 + O(\lambda^{-1}))e(\lambda)^{-1}. \tag{6.13}$$

43

The coefficient $Q5$ before $\mathcal{E}(\cdot)$ can be approximated as

$$Q5 \equiv \left[-EI(\frac{\hat{\gamma}(\lambda)}{\gamma(\lambda)})^2 + g_h\lambda\frac{\hat{\gamma}(\lambda)}{\gamma^2(\lambda)} \right] \hat{e}(\lambda)^{-1} = [-EI + g_hP^{-1}\lambda^{1/2}](1 + O(\lambda^{-1}))\hat{e}(\lambda)^{-1}.$$

(6.14)

The coefficient $Q6$ before $\mathcal{F}(\cdot)$ can be approximated as

$$Q6 \equiv \left[-EI(\frac{\Gamma(\lambda)}{\gamma(\lambda)})^2 + g_h\lambda\frac{\Gamma(\lambda)}{\gamma^2(\lambda)} \right] e_+(\lambda)^{-1}$$

$$= [-EIQ^2P^{-2}\lambda + g_h\lambda QP^{-2}](1 + O(\lambda^{-1}))e_+(\lambda)^{-1}.$$

(6.15)

Thus we can write the second right–hand boundary condition by summing the results of (6.10)–(6.15) and setting the sum equal to zero. Again we leave the terms involving $\mathcal{A}(\cdot)$, $\mathcal{B}(\cdot)$ and $\mathcal{D}(\cdot)$ on the left side and take the other terms to the right side to rewrite Eq.(6.8) in the form

$$Q1\,\mathcal{A}(\lambda) + Q2\,\mathcal{B}(\lambda) + Q3\,\mathcal{C}(\lambda) = -\left[Q4\,\mathcal{D}(\lambda) + Q5\,\mathcal{E}(\lambda) + Q6\,\mathcal{F}(\lambda) \right].$$

The latter equation has the following asymptotic approximation:

$$\left[EI + g_h\frac{1}{P}i\lambda^{1/2} \right](1 + O(\lambda^{-1}))\mathcal{A}(\lambda)e(\lambda) + \left[-EI - g_h\frac{1}{P}\lambda^{1/2} \right] \times$$

$$(1 + O(\lambda^{-1}))\mathcal{B}(\lambda)\hat{e}(\lambda) - \left[EI\frac{Q^2}{P^2}\lambda + g_h\lambda\frac{Q}{P^2} \right](1 + O(\lambda^{-1}))\mathcal{C}(\lambda)e_+(\lambda) =$$

$$\left[-EI + g_h\frac{1}{P}i\lambda^{1/2} \right](1 + O(\lambda^{-1}))\mathcal{D}(\lambda)e(\lambda)^{-1} + \left[EI - g_h\frac{1}{P}\lambda^{1/2} \right] \times$$

$$(1 + O(\lambda^{-1}))\mathcal{E}(\lambda)\hat{e}(\lambda)^{-1} + \left[EI\frac{Q^2}{P^2}\lambda - g_h\lambda\frac{Q}{P^2} \right](1 + O(\lambda^{-1}))\mathcal{F}(\lambda)e_+(\lambda)^{-1}.$$

(6.16)

Examining the third right–hand boundary condition given in (3.24), we realize that we need the function $\Psi(\cdot)$, along with its first, fourth, and fifth derivatives evaluated at zero. So using $\Psi(\cdot)$ from (4.87) and definition (6.1) yields

$$\Psi(\lambda, 0) = \mathcal{A}(\lambda)e(\lambda) + \mathcal{B}(\lambda)\hat{e}(\lambda) + \mathcal{C}(\lambda)e_+(\lambda) + \mathcal{D}(\lambda)e(\lambda)^{-1} + \mathcal{E}(\lambda)\hat{e}(\lambda)^{-1} + \mathcal{F}(\lambda)e_+(\lambda)^{-1}.$$

(6.17)

The first derivative of the general solution evaluated at $x = 0$ is

$$\Psi'(\lambda, 0) = \gamma(\lambda)\mathcal{A}(\lambda)e(\lambda) + i\hat{\gamma}(\lambda)\mathcal{B}(\lambda)\hat{e}(\lambda) + i\Gamma(\lambda)\mathcal{C}(\lambda)e_+(\lambda) -$$

$$\gamma(\lambda)\mathcal{D}(\lambda)e(\lambda)^{-1} - i\hat{\gamma}(\lambda)\mathcal{E}(\lambda)\hat{e}(\lambda)^{-1} - i\Gamma(\lambda)\mathcal{F}(\lambda)e_+(\lambda)^{-1}.$$

(6.18)

44

The fourth derivative of the function $\Psi(\cdot)$ evaluated at $x = 0$ is

$$\Psi''''(\lambda, 0) = \gamma^4(\lambda)\mathcal{A}(\lambda)e(\lambda) + \hat{\gamma}^4(\lambda)\mathcal{B}(\lambda)\hat{e}(\lambda) + \Gamma^4(\lambda)\mathcal{C}(\lambda)e_+(\lambda) +$$
$$\gamma^4(\lambda)\mathcal{D}(\lambda)e(\lambda)^{-1} + \hat{\gamma}^4(\lambda)\mathcal{E}(\lambda)\hat{e}(\lambda)^{-1} + \Gamma^4(\lambda)\mathcal{F}(\lambda)e_+(\lambda)^{-1}. \tag{6.19}$$

And the fifth derivative of the function $\Psi(\cdot)$ evaluated at $x = 0$ is

$$\Psi^V(\lambda, 0) = \gamma^5(\lambda)\mathcal{A}(\lambda)e(\lambda) + i\hat{\gamma}^5(\lambda)\mathcal{B}(\lambda)\hat{e}(\lambda) + i\Gamma^5(\lambda)\mathcal{C}(\lambda)e_+(\lambda) -$$
$$\gamma^5(\lambda)\mathcal{D}(\lambda)e(\lambda)^{-1} - i\hat{\gamma}^5(\lambda)\mathcal{E}(\lambda)\hat{e}(\lambda)^{-1} - i\Gamma^5(\lambda)\mathcal{F}(\lambda)e_+(\lambda)^{-1}. \tag{6.20}$$

After substituting results (6.18)–(6.20) into the sixth boundary condition (3.24), we divide the resulting equation by $\gamma^2(\cdot)$. The following approximations are valid:

$$\frac{\hat{\gamma}^4(\lambda)}{\gamma^2(\lambda)} = P^2\lambda(1 + O(\lambda^{-1})), \quad \frac{1}{\gamma^2(\lambda)} = \frac{1}{P^2}\lambda^{-1}(1 + O(\lambda^{-1})),$$
$$\frac{\Gamma^5(\lambda)}{\gamma^2(\lambda)} = \frac{Q^5}{P^2}\lambda^4(1 + O(\lambda^{-1})), \quad \frac{\Gamma^4(\lambda)}{\gamma^2(\lambda)} = \frac{Q^4}{P^2}\lambda^3(1 + O(\lambda^{-1})), \tag{6.21}$$

where P and Q are defined in (4.86). Now we substitute (6.17)–(6.20) into the third

right–hand boundary condition (3.24) leaving the terms involving $\mathcal{A}(\cdot)$, $\mathcal{B}(\cdot)$ and $\mathcal{C}(\cdot)$ on the left side, and moving the terms involving $\mathcal{D}(\cdot)$, $\mathcal{E}(\cdot)$ and $\mathcal{F}(\cdot)$ to the right side.

Let us collect together all the terms involving $\mathcal{A}(\cdot)e(\cdot)$ and denote the coefficient for $\mathcal{A}(\cdot)e(\cdot)$ by $\mathbb{A}_{31}$. For $\mathbb{A}_{31}$, we have

$$\mathbb{A}_{31}(\lambda) = EI\, GJ\gamma^3(\lambda) + iEI\, g_\alpha\lambda\gamma^2(\lambda) - m\, GJ\, \lambda^2 - ig_\alpha m\lambda^3\gamma^{-2}(\lambda)$$
$$= \left[EI\, GJP^3\lambda^{3/2} + iEI\, g_\alpha P^2\lambda^2 - m\, GJ\frac{1}{P}\lambda^{3/2} - ig_\alpha m\frac{1}{P^2}\lambda^2 \right](1 + O(\lambda^{-1/2}))$$
$$= i\left[EI\, g_\alpha P^2 - g_\alpha m\frac{1}{P^2} \right]\lambda^2(1 + O(\lambda^{-1/2})). \tag{6.22}$$

Let us collect together all terms involving $\mathcal{D}(\cdot)e(\cdot)^{-1}$ and denote them by $\mathbb{B}_{31}$. For $\mathbb{B}_{31}$, we have

$$\mathbb{B}_{31}(\lambda) = EI\, GJ\gamma^3(\lambda) - iEI\, g_\alpha\lambda\gamma^2(\lambda) - m\, GJ\, \lambda^2 + ig_\alpha m\lambda^3\gamma^{-2}(\lambda)$$
$$= i\left[-EI\, g_\alpha P^2 + g_\alpha m\frac{1}{P^2} \right]\lambda^2(1 + O(\lambda^{-1/2})). \tag{6.23}$$

45

Let us collect together all the terms involving $\mathcal{B}(\cdot)\hat{e}(\cdot)$ and denote them by $\mathbb{A}_{32}$. For $\mathbb{A}_{32}$, we have

$$
\begin{aligned}
\mathbb{A}_{32}(\lambda) = {} & iEI\,GJ\hat{\gamma}^5(\lambda)\gamma^{-2}(\lambda) + iEI\,g_\alpha\lambda\hat{\gamma}^4(\lambda)\gamma^{-2}(\lambda) - im\,GJ\,\lambda^2\hat{\gamma}(\lambda)\gamma^{-2}(\lambda) \\
& - ig_\alpha m\lambda^3\gamma^{-2}(\lambda) = \left[iEI\,g_\alpha\lambda P^2\lambda - ig_\alpha m\lambda^3\frac{1}{P^2}\lambda^{-1} \right](1 + O(\lambda^{-1/2})) \\
= {} & i\left[EI\,g_\alpha P^2 - g_\alpha m\frac{1}{P^2} \right]\lambda^2(1 + O(\lambda^{-1/2})).
\end{aligned}
\tag{6.24}
$$

Let us collect together all the terms involving $\mathcal{E}(\cdot)\hat{e}(\cdot)^{-1}$ and denote them by $\mathbb{B}_{32}$. For $\mathbb{B}_{32}$, we have

$$
\begin{aligned}
\mathbb{B}_{32}(\lambda) = {} & iEI\,GJ\hat{\gamma}^5(\lambda)\gamma^{-2}(\lambda) - iEI\,g_\alpha\lambda\hat{\gamma}^4(\lambda)\gamma^{-2}(\lambda) - im\,GJ\,\lambda^2\hat{\gamma}(\lambda)\gamma^{-2}(\lambda) \\
& + ig_\alpha m\lambda^3\gamma^{-2}(\lambda) = \left[-iEI\,g_\alpha\lambda P^2\lambda + ig_\alpha m\lambda^3\frac{1}{P^2}\lambda^{-1} \right](1 + O(\lambda^{-1/2})) \\
= {} & i\left[-EI\,g_\alpha P^2 + g_\alpha m\frac{1}{P^2} \right]\lambda^2(1 + O(\lambda^{-1/2})).
\end{aligned}
$$

$$\tag{6.25}$$

Let us collect together all the terms involving $\mathcal{C}(\cdot)e_+(\cdot)$ and denote them by $\mathbb{A}_{33}$. For $\mathbb{A}_{33}$, we have

$$
\begin{aligned}
\mathbb{A}_{33}(\lambda) = {} & iEI\,GJ\Gamma^5(\lambda)\gamma^{-2}(\lambda) + iEI\,g_\alpha\lambda\Gamma^4(\lambda)\gamma^{-2}(\lambda) - im\,GJ\,\lambda^2\Gamma(\lambda)\gamma^{-2}(\lambda) \\
& - ig_\alpha m\lambda^3\gamma^{-2}(\lambda) = iEI\,GJ\frac{Q^5}{P^2}\lambda^4(1 + O(\lambda^{-1})) + iEI\,g_\alpha\lambda\frac{Q^4}{P^2}\lambda^3(1 + O(\lambda^{-1})) \\
= {} & i\left[EI\,GJ\frac{Q^5}{P^2} + EIg_\alpha\frac{Q^4}{P^2} \right]\lambda^4(1 + O(\lambda^{-1})).
\end{aligned}
$$

$$\tag{6.26}$$

Let us collect together all the terms involving $\mathcal{F}(\cdot)e_+(\cdot)^{-1}$ and denote them by $\mathbb{B}_{33}$. For $\mathbb{B}_{33}$, we have

$$
\begin{aligned}
\mathbb{B}_{33}(\lambda) = {} & iEI\,GJ\Gamma^5(\lambda)\gamma^{-2}(\lambda) - iEI\,g_\alpha\lambda\Gamma^4(\lambda)\gamma^{-2}(\lambda) - im\,GJ\,\lambda^2\Gamma(\lambda)\gamma^{-2}(\lambda) \\
& + ig_\alpha m\lambda^3\gamma^{-2}(\lambda) = iEI\,GJ\frac{Q^5}{P^2}\lambda^4(1 + O(\lambda^{-1})) - iEI\,g_\alpha\lambda\frac{Q^4}{P^2}\lambda^3(1 + O(\lambda^{-1})) \\
= {} & i\left[EI\,GJ\frac{Q^5}{P^2} - EIg_\alpha\frac{Q^4}{P^2} \right]\lambda^4(1 + O(\lambda^{-1})).
\end{aligned}
$$

$$\tag{6.27}$$

Thus, we can write the third boundary condition (3.24) in the form

$$
\mathbb{A}_{31}(\lambda)\mathcal{A}(\lambda)e(\lambda) + \mathbb{A}_{32}(\lambda)\mathcal{B}(\lambda)\hat{e}(\lambda) + \mathbb{A}_{33}(\lambda)\mathcal{C}(\lambda)e_+(\lambda) =
$$
$$
\mathbb{B}_{31}(\lambda)\mathcal{D}(\lambda)e(\lambda)^{-1} + \mathbb{B}_{32}(\lambda)\mathcal{E}(\lambda)\hat{e}(\lambda)^{-1} + \mathbb{B}_{33}(\lambda)\mathcal{F}(\lambda)e_+(\lambda)^{-1},
$$
(6.28)

where the coefficients $\mathbb{A}_{ij}$, $\mathbb{B}_{ij}$, $i = 3$, $j = 1, 2, 3$, are given by formulae (6.22)–(6.27). Setting

$$
\mathbb{E}(\lambda) = diag(e(\lambda), \hat{e}(\lambda), e_+(\lambda)),
$$
(6.29)

we can write the three right–hand boundary conditions in the matrix form. This matrix form, which will appear below, contains two matrices denoted by $\mathbb{A}(\cdot)$ and $\mathbb{B}(\cdot)$. They are similar to the ones appearing in formula (5.18). Without any misunderstanding, we will use the notation keeping in mind that the new matrices $\mathbb{A}(\cdot)$ and $\mathbb{B}(\cdot)$ are different from the ones which appeared in Section 5.2. Thus, we have

$$
\mathbb{A}(\lambda)\,\mathbb{E}(\lambda)\,X(\lambda) = \mathbb{B}(\lambda)\,\mathbb{E}^{-1}(\lambda)\,Y(\lambda),
$$
(6.30)

where the entries of the matrices of $\mathbb{A}(\cdot)$ and $\mathbb{B}(\cdot)$ are given in (6.6), (6.16), and (6.22)–(6.28).

To rewrite the matrix equation (6.30) explicitly, it is convenient to introduce the following notation: $\omega_{ij}(\lambda)$ and $\hat{\omega}_{ij}(\lambda)$; $\omega_{ij}(\lambda)$, with $i, j = 1, 2, 3$, means that there is a factor $(1 + O(\lambda^{-1/2}))$ on the intersection of the i-th row and the j-th column in the matrix below, and $\hat{\omega}_{ij}(\lambda)$, with $i, j = 1, 2, 3$, means that there is a factor $(1 + O(\lambda^{-1}))$ on the intersection of the i-th row and the j-th column in the matrix below. With the aforementioned notation, the matrix equation (6.30) then becomes

$$\begin{bmatrix} 1 & -i\hat{\omega}_{12}(\lambda) & i\frac{Q^3}{P^3}\lambda^{3/2}\hat{\omega}_{13}(\lambda) \\ Q1(\lambda)e(\lambda)^{-1} & Q2(\lambda)\hat{e}(\lambda)^{-1} & Q3(\lambda)e_+(\lambda)^{-1} \\ \mathbb{A}_{31}(\lambda) & \mathbb{A}_{32}(\lambda) & \mathbb{A}_{33}(\lambda) \end{bmatrix} \mathbb{E}(\lambda)X(\lambda) =$$

$$(6.31)$$

$$\begin{bmatrix} 1 & -i\hat{\omega}_{12}(\lambda) & -i\frac{Q^3}{P^3}\lambda^{3/2}\hat{\omega}_{13}(\lambda) \\ -Q4(\lambda)e(\lambda) & -Q5(\lambda)\hat{e}(\lambda) & -Q6(\lambda)e_+(\lambda) \\ -\mathbb{B}_{31}(\lambda) & -\mathbb{B}_{32}(\lambda) & -\mathbb{B}_{33}(\lambda) \end{bmatrix} \mathbb{E}^{-1}(\lambda)Y(\lambda),$$

where $Q1$, $Q2$, $Q3$, $Q4$, $Q5$, and $Q6$ are defined by formulae (6.10)–(6.15); the entries $\mathbb{A}_{3i}(\lambda)$, $i = 1, 2, 3$, are defined by formulae (6.22), (6.24), and (6.26); the entries $\mathbb{B}_{3i}(\lambda)$, $i = 1, 2, 3$, are defined by formulae (6.23), (6.25), and (6.27) respectively.

Assuming that the matrix $\mathbb{A}^{-1}(\cdot)$ exists, we solve (6.30) for $X(\cdot)$ and obtain

$$X(\lambda) = \mathbb{E}^{-1}(\lambda)\mathbb{A}^{-1}(\lambda)\mathbb{B}(\lambda)\mathbb{E}^{-1}(\lambda)Y(\lambda). \tag{6.32}$$

Comparing (5.1) with (6.32), we conclude that the right–reflection matrix can be represented as

$$\mathbb{R}_r(\lambda) = \mathbb{E}^{-1}(\lambda)\mathbb{A}^{-1}(\lambda)\mathbb{B}(\lambda)\mathbb{E}^{-1}(\lambda). \tag{6.33}$$

It is this right–reflection matrix that will be calculated in the remainder of this section. While we could certainly compute $\mathbb{A}(\cdot)^{-1}\mathbb{B}(\cdot)$ in a straightforward manner, we will instead notice that $\mathbb{B}(\cdot)$ can be thought of as a "perturbation" of $\mathbb{A}(\cdot)$, i.e., let

$$\mathbb{B}(\lambda) = \mathbb{A}(\lambda) - \mathbb{V}(\lambda). \tag{6.34}$$

For $\mathbb{V}(\cdot)$, we obtain the expression

48

$$\mathbb{V}(\lambda) = 2 \begin{bmatrix} 1 & 0 & 0 \\ 0 & EI & 0 \\ 0 & 0 & i\lambda^2 g_\alpha \end{bmatrix} \times$$

$$\begin{bmatrix} 0 & 0 & 0 \\ \omega_{21}(\lambda) & -\omega_{22}(\lambda) & -\lambda\dfrac{Q^2}{P^2}\omega_{23}(\lambda) \\ \left[EI\,P^2 - \dfrac{m}{P^2}\right]\omega_{31}(\lambda) & \left[EI\,P^2 - \dfrac{m}{P^2}\right]\omega_{32}(\lambda) & EI\dfrac{Q^4}{P^4}\lambda^2\omega_{33}(\lambda) \end{bmatrix} \tag{6.35}$$

Thus the matrix $\mathbb{A}(\cdot)^{-1}\mathbb{B}(\cdot)$ can be written as

$$\mathbb{A}(\lambda)^{-1}\mathbb{B}(\lambda) = \mathbb{A}(\lambda)^{-1}(\mathbb{A}(\lambda) - \mathbb{V}(\lambda)) = \mathbb{I} - \mathbb{A}(\lambda)^{-1}\mathbb{V}(\lambda). \tag{6.36}$$

Since $\mathbb{V}(\cdot)$ has a row of zeros, it will be more efficient to calculate $\mathbb{A}(\lambda)^{-1}\mathbb{V}(\lambda)$ instead of $\mathbb{A}(\lambda)^{-1}\mathbb{B}(\lambda)$.

Now we begin to calculate an asymptotic representation for $\mathbb{A}(\lambda)^{-1}$. From (6.31), we derive that

$$\mathbb{A}(\lambda) = \begin{bmatrix} 1 & -i\hat{\omega}_{12}(\lambda) & -i\lambda^{1/2}\dfrac{Q^3}{P^2}\hat{\omega}_{13}(\lambda) \\ i\lambda^{1/2}\dfrac{gh}{P}\omega_{21}(\lambda) & -\lambda^{1/2}\dfrac{gh}{P}\omega_{22}(\lambda) & -\lambda\dfrac{Q}{P^2}\left[EIQ + g_h\right]\hat{\omega}_{32}(\lambda) \\ i\left[EI\,P^2 g_\alpha - \dfrac{mg_\alpha}{P^2}\right]\lambda^2\omega_{31}(\lambda) & i\lambda^2\left[EI\,P^2 g_\alpha - \dfrac{mg_\alpha}{P^2}\right]\omega_{32}(\lambda) & iEI\dfrac{Q^4}{P^4}\left[GJQ + g_\alpha\right]\lambda^4\hat{\omega}_{33}(\lambda) \end{bmatrix} \tag{6.37}$$

Setting

$$\mathbb{A}(\lambda) \equiv \begin{bmatrix} 1 & -i\hat{\omega}_{12}(\lambda) & a_{13}\lambda^{3/2}\hat{\omega}_{13}(\lambda) \\ a_{21}\lambda^{1/2}\omega_{21}(\lambda) & a_{22}\lambda^{1/2}\omega_{22}(\lambda) & a_{23}\lambda\hat{\omega}_{23}(\lambda) \\ a_{31}\lambda^2\omega_{31}(\lambda) & a_{32}\lambda^2\omega_{32}(\lambda) & a_{33}\lambda^4\hat{\omega}_{33}(\lambda) \end{bmatrix}, \tag{6.38}$$

we calculate an asymptotic approximation to $det\,\mathbb{A}(\lambda)$ by expanding along the entries of the third row and noting that the highest power of λ comes from the term with a_{33}

$$\begin{aligned} det(\mathbb{A}(\lambda)) &= a_{33}\lambda^4\hat{\omega}_{33}(\lambda)[a_{22}\lambda^{1/2}(1+O(\lambda^{-1/2})) + a_{21}i\lambda^{1/2}(1+O(\lambda^{-1/2}))] \\ &= a_{33}(a_{22}+ia_{21})\lambda^{9/2}(1+O(\lambda^{-1/2})). \end{aligned} \tag{6.39}$$

Next we calculate each entry of $\mathbb{A}(\cdot)^{-1}$ by dividing the appropriate cofactor by $det\,\mathbb{A}(\lambda)$. Beginning with the top row, we calculate $(\mathbb{A}(\lambda)^{-1})_{11}$

$$\begin{aligned} (\mathbb{A}(\lambda)^{-1})_{11} &= \frac{|\mathbb{A}(\lambda)|_{11}}{det\,\mathbb{A}(\lambda)} = \frac{a_{22}a_{33}\lambda^{9/2}(1+O(\lambda^{-1/2})) - a_{23}a_{32}\lambda^3(1+O(\lambda^{-1/2}))}{a_{33}(a_{22}+ia_{21})\lambda^{9/2}(1+O(\lambda^{-1/2}))} \\ &= \frac{a_{22}}{a_{22}+ia_{21}}(1+O(\lambda^{-1/2})), \end{aligned} \tag{6.40}$$

where $|\mathbb{A}(\cdot)|_{11}$ is a cofactor of $\mathbb{A}(\cdot)$ corresponding to a_{11}. Note that in (6.40), we have kept the entire formal result, and have not deleted the second term even though it is not within the accuracy of the asymptotic calculations. Now we proceed to find the remaining entries in the first row. The second entry is

$$\begin{aligned} (\mathbb{A}(\lambda)^{-1})_{12} &= -\frac{|\mathbb{A}(\lambda)|_{21}}{det\,\mathbb{A}(\lambda)} = \frac{ia_{33}\lambda^4 + a_{32}\lambda^2 a_{13}\lambda^{3/2}(1+O(\lambda^{-1}))}{a_{33}(a_{22}+ia_{21})\lambda^{9/2}(1+O(\lambda^{-1/2}))} \\ &= \frac{i}{a_{22}+ia_{21}}\lambda^{-1/2}(1+O(\lambda^{-1/2})), \end{aligned} \tag{6.41}$$

and the third entry of the first row is calculated as

$$(\mathbb{A}(\lambda)^{-1})_{13} = \frac{|\mathbb{A}(\lambda)|_{31}}{det\,\mathbb{A}(\lambda)} = \frac{-ia_{23}\lambda - a_{22}\lambda^{1/2}(1+O(\lambda^{-1/2}))a_{13}\lambda^{3/2}}{a_{33}(a_{22}+ia_{21})\lambda^{9/2}(1+O(\lambda^{-1/2}))} = O(\lambda^{-5/2}). \tag{6.42}$$

Now we move to the second row of $\mathbb{A}(\lambda)^{-1}$, whose first entry is

$$
\begin{aligned}
(\mathbb{A}(\lambda)^{-1})_{21} &= -\frac{|\mathbb{A}(\lambda)|_{12}}{\det \mathbb{A}(\lambda)} = \frac{-[a_{21}\lambda^{1/2}a_{33}\lambda^4 - a_{31}a_{23}\lambda^2\lambda(1+O(\lambda^{-1/2}))]}{a_{33}(a_{22}+ia_{21})\lambda^{9/2}(1+O(\lambda^{-1/2}))} \\
&= -\frac{a_{21}}{(a_{22}+ia_{21})}(1+O(\lambda^{-1/2})).
\end{aligned}
\tag{6.43}
$$

The second entry is

$$
(\mathbb{A}(\lambda)^{-1})_{22} = \frac{|\mathbb{A}(\lambda)|_{22}}{\det \mathbb{A}(\lambda)} = \frac{a_{33}\lambda^4 - a_{13}a_{31}\lambda^{7/2}(1+O(\lambda^{-1/2}))}{a_{33}(a_{22}+ia_{21})\lambda^{9/2}(1+O(\lambda^{-1/2}))} = O(\lambda^{-1/2}).
\tag{6.44}
$$

The third entry is

$$
(\mathbb{A}(\lambda)^{-1})_{23} = -\frac{|\mathbb{A}(\lambda)|_{32}}{\det \mathbb{A}(\lambda)} = \frac{a_{23}\lambda - a_{21}a_{13}\lambda^2(1+O(\lambda^{-1/2}))}{a_{33}(a_{22}+ia_{21})\lambda^{9/2}(1+O(\lambda^{-1/2}))} = O(\lambda^{-5/2}).
\tag{6.45}
$$

Finally we turn to the third row, whose first entry is

$$
(\mathbb{A}(\lambda)^{-1})_{31} = \frac{|\mathbb{A}(\lambda)|_{13}}{\det \mathbb{A}(\lambda)} = \frac{[a_{21}a_{32}\lambda^{5/2} - a_{31}a_{22}\lambda^{5/2}](1+O(\lambda^{-1/2}))}{a_{33}(a_{22}+ia_{21})\lambda^{9/2}(1+O(\lambda^{-1/2}))} = O(\lambda^{-2}).
\tag{6.46}
$$

The second entry is

$$
(\mathbb{A}(\lambda)^{-1})_{32} = -\frac{|\mathbb{A}(\lambda)|_{23}}{\det \mathbb{A}(\lambda)} = -\frac{[a_{32}\lambda^2 + ia_{31}\lambda^2](1+O(\lambda^{-1/2}))}{a_{33}(a_{22}+ia_{21})\lambda^{9/2}(1+O(\lambda^{-1/2}))} = O(\lambda^{-5/2}).
\tag{6.47}
$$

The third entry is

$$
(\mathbb{A}(\lambda)^{-1})_{33} = \frac{|\mathbb{A}(\lambda)|_{33}}{\det \mathbb{A}(\lambda)} = \frac{a_{22}\lambda^{1/2} + ia_{21}\lambda^{1/2}(1+O(\lambda^{-1/2}))}{a_{33}(a_{22}+ia_{21})\lambda^{9/2}(1+O(\lambda^{-1/2}))} = O(\lambda^{-4}).
\tag{6.48}
$$

Now we express the entries of $\mathbb{A}(\lambda)^{-1}$ in terms of the original parameters. We return to (6.37) and (6.38) to calculate that

$$
a_{22} + ia_{21} = -2g_h P^{-1}.
\tag{6.49}
$$

Now we can return to the results of (6.40)–(6.48) to calculate each individual entry of $\mathbb{A}(\lambda)^{-1}$. We find that (6.40) becomes

$$
(\mathbb{A}(\lambda)^{-1})_{11} = \frac{1}{2}\left(1+O(\lambda^{-1/2})\right).
\tag{6.50}
$$

We find that (6.41) and (6.42) become

$$
(\mathbb{A}(\lambda)^{-1})_{12} = O(\lambda^{-1/2}), \qquad (\mathbb{A}(\lambda)^{-1})_{13} = O(\lambda^{-5/2}).
\tag{6.51}
$$

51

We find that (6.43) becomes

$$(\mathbb{A}(\lambda)^{-1})_{21} = \frac{-a_{21}}{a_{22} + ia_{21}}\omega(\lambda) = \frac{i}{2}\left(1 + O(\lambda^{-1/2})\right). \tag{6.52}$$

It can be verified that the asymptotic precisions of the remaining terms are

$$(\mathbb{A}(\lambda)^{-1})_{22} = O(\lambda^{-1/2}), \quad (\mathbb{A}(\lambda)^{-1})_{23} = O(\lambda^{-5/2}), \quad (\mathbb{A}(\lambda)^{-1})_{31} = O(\lambda^{-2}),$$

$$(\mathbb{A}(\lambda)^{-1})_{32} = O(\lambda^{-5/2}), \quad (\mathbb{A}(\lambda)^{-1})_{33} = O(\lambda^{-4}). \tag{6.53}$$

Taking into account (6.50)–(6.53), we obtain the following representation:

$$\mathbb{A}(\lambda)^{-1} = \begin{bmatrix} \frac{1}{2}(1 + O(\lambda^{-1/2})) & O(\lambda^{-1/2}) & O(\lambda^{-5/2}) \\[2ex] \frac{i}{2}(1 + O(\lambda^{-1/2})) & O(\lambda^{-1/2}) & O(\lambda^{-5/2}) \\[2ex] O(\lambda^{-2}) & O(\lambda^{-5/2}) & O(\lambda^{-4}) \end{bmatrix}. \tag{6.54}$$

Now we are in a position to compute the right–reflection matrix. Using (6.33) and (6.36), we recall that

$$\mathbb{R}_r(\lambda) = \mathbb{E}^{-1}(\lambda)\left[\mathbb{I} - \mathbb{A}(\lambda)^{-1}\mathbb{V}(\lambda)\right]\mathbb{E}^{-1}(\lambda), \tag{6.55}$$

where $\mathbb{A}(\lambda)^{-1}$ is given by (6.54) and

$$\mathbb{V}(\lambda) = \begin{bmatrix} 0 & 0 & 0 \\[2ex] 2EI\hat{\omega}_{21}(\lambda) & -2EI\hat{\omega}_{22}(\lambda) & -2EI\dfrac{Q^2}{P^2}\lambda\hat{\omega}_{23}(\lambda) \\[2ex] 2v_{31}\lambda^2\omega_{31}(\lambda) & 2v_{32}\lambda^2\omega_{32}(\lambda) & 2v_{32}\lambda^4\hat{\omega}_{33}(\lambda) \end{bmatrix}, \tag{6.56}$$

where

$$v_{31} \equiv (\mathbb{V}(\lambda))_{31} = iEI\, g_\alpha P^2 - ig_\alpha mP^{-2},$$

$$v_{32} \equiv (\mathbb{V}(\lambda))_{32} = iEI\, g_\alpha P^2 - ig_\alpha mP^{-2} = (\mathbb{V}(\lambda))_{31}, \tag{6.57}$$

$$v_{33} \equiv (\mathbb{V}(\lambda))_{33} = iEI\, g_\alpha Q^4 P^{-2}.$$

52

We compute $\mathbb{A}(\lambda)^{-1}\mathbb{V}(\lambda)$ as

$$\mathbb{A}(\lambda)^{-1}\mathbb{V}(\lambda) = \begin{bmatrix} O(\lambda^{-1/2}) & O(\lambda^{-1/2}) & r_{13}\lambda^{3/2}\left(1+O(\lambda^{-1})\right) \\ \\ O(\lambda^{-1/2}) & O(\lambda^{-1/2}) & r_{23}\lambda^{3/2}\left(1+O(\lambda^{-1})\right) \\ \\ O(\lambda^{-2}) & O(\lambda^{-2}) & r_{33}\left(1+O(\lambda^{-1})\right) \end{bmatrix}, \qquad (6.58)$$

where the precise values of the constants r_{13} and r_{23} are not important for us. However, the expression for r_{33} is crucially important. We now calculate r_{33} and have (see (6.48))

$$(\mathbb{A}(\lambda)^{-1})_{33} = \frac{1}{a_{33}}\lambda^{-4}(1+O(\lambda^{-1/2})) = \frac{P^2}{iQ^4 EI\left[GJ\,Q+g_\alpha\right]}\lambda^{-4}(1+O(\lambda^{-1/2})), \;\; (6.59)$$

which we then use to find that

$$\begin{aligned} r_{33}(1+O(\lambda^{-1})) &= \mathbb{A}(\lambda)_{33}^{-1}2v_{33}\lambda^4(1+O(\lambda^{-1})) \\ &= \frac{2EI\,g_\alpha Q^4(1+O(\lambda^{-1/2}))}{[EI\,GJQ^5+EI\,g_\alpha Q^4]} = \frac{2g_\alpha(1+O(\lambda^{-1/2}))}{GJQ+g_\alpha} \\ &= \frac{2g_\alpha(1+O(\lambda^{-1/2}))}{(I_\alpha GJ)^{1/2}+g_\alpha}. \end{aligned} \qquad (6.60)$$

Using the latter result, we calculate that

$$\mathbb{I}-\mathbb{A}(\lambda)^{-1}\mathbb{V}(\lambda) = \begin{bmatrix} \left(1+O(\lambda^{-1/2})\right) & O(\lambda^{-1/2}) & -r_{13}\lambda^{3/2}\left(1+O(\lambda^{-1})\right) \\ \\ O(\lambda^{-1/2}) & \left(1+O(\lambda^{-1/2})\right) & -r_{23}\lambda^{3/2}\left(1+O(\lambda^{-1})\right) \\ \\ O(\lambda^{-2}) & O(\lambda^{-2}) & \left(1-r_{33}\right)\left(1+O(\lambda^{-1})\right) \end{bmatrix}.$$

$$(6.61)$$

Now we are in a position to calculate the right–reflection matrix. We have

$$\mathbb{R}_r = \mathbb{E}^{-1}(\lambda)(\mathbb{I} - \mathbb{A}(\lambda)^{-1}\mathbb{V}(\lambda))\mathbb{E}^{-1}(\lambda)$$

$$= \begin{bmatrix} e(\lambda)^{-1} & 0 & 0 \\ 0 & \hat{e}(\lambda)^{-1} & 0 \\ 0 & 0 & e_+(\lambda)^{-1} \end{bmatrix} \begin{bmatrix} \left(1+O(\lambda^{-1/2})\right) & O(\lambda^{-1/2}) & -r_{13}\lambda^{3/2}\left(1+O(\lambda^{-1})\right) \\ O(\lambda^{-1/2}) & \left(1+O(\lambda^{-1/2})\right) & -r_{23}\lambda^{3/2}\left(1+O(\lambda^{-1})\right) \\ O(\lambda^{-2}) & O(\lambda^{-2}) & (1-r_{33})\left(1+O(\lambda^{-1})\right) \end{bmatrix} \times$$

$$\begin{bmatrix} e(\lambda)^{-1} & 0 & 0 \\ 0 & \hat{e}(\lambda)^{-1} & 0 \\ 0 & 0 & e_+(\lambda)^{-1} \end{bmatrix}$$

$$(6.62)$$

$$\mathbb{R}_r = \begin{bmatrix} e(\lambda)^{-2}(1+O(\lambda^{-1/2})) & [e(\lambda)\hat{e}(\lambda)]^{-1}O(\lambda^{-1/2}) & -[e(\lambda)e_+(\lambda)]^{-1}r_{13}\lambda^{3/2}(1+O(\lambda^{-1})) \\ [e(\lambda)\hat{e}(\lambda)]^{-1}O(\lambda^{-1/2}) & \hat{e}(\lambda)^{-2}(1+O(\lambda^{-1/2})) & -[\hat{e}(\lambda)e_+(\lambda)]^{-1}r_{23}\lambda^{3/2}(1+O(\lambda^{-1})) \\ [e(\lambda)e_+(\lambda)]^{-1}O(\lambda^{-2}) & [\hat{e}(\lambda)e_+(\lambda)]^{-1}O(\lambda^{-2}) & e_+(\lambda)^{-2}(1-r_{33})(1+O(\lambda^{-1})) \end{bmatrix}$$

$$(6.63)$$

Thus, computation of the right–reflection matrix $\mathbb{R}_r$ is complete.

CHAPTER VII

Spectral Asymptotics

7.1 Spectral Equation.

In this section, we are in a position to give an asymptotic form for the equation, whose solutions will give us asymptotic representations for the spectrum. We reproduce the main equation from Section 5.1

$$\det(\mathbb{R}_l(\lambda) - \mathbb{R}_r(\lambda)) = 0. \tag{7.1}$$

Using asymptotic approximations for the reflection matrices from Chapters V and VI (see formulae (5.51) and (6.63)), we have

$$\mathbb{R}_l(\lambda) - \mathbb{R}_r(\lambda) = \begin{bmatrix} i\hat{\omega}_{11}(\lambda) & (i-1)\hat{\omega}_{12}(\lambda) & \sqrt{\dfrac{I_\alpha}{GJ}}P^{-1}(i-1)\lambda^{1/2}\hat{\omega}_{13}(\lambda) \\[2em] -(i+1)\hat{\omega}_{21}(\lambda) & -i\hat{\omega}_{22}(\lambda) & \sqrt{\dfrac{I_\alpha}{GJ}}P^{-1}(1-i)\lambda^{1/2}\hat{\omega}_{32}(\lambda) \\[2em] O(\lambda^{-3}) & O(\lambda^{-3}) & -1+O(\lambda^{-2.5}) \end{bmatrix} -$$

$$\begin{bmatrix} e(\lambda)^{-2}\omega_{11}(\lambda) & [e(\lambda)\hat{e}(\lambda)]^{-1}O(\lambda^{-1/2}) & -[e(\lambda)e_+(\lambda)]^{-1}r_{13}\lambda^{3/2}\hat{\omega}_{13}(\lambda) \\[2em] [e(\lambda)\hat{e}(\lambda)]^{-1}O(\lambda^{-1/2}) & \hat{e}(\lambda)^{-2}\omega_{22}(\lambda) & -[\hat{e}(\lambda)e_+(\lambda)]^{-1}r_{23}\lambda^{3/2}\hat{\omega}_{23}(\lambda) \\[2em] [e(\lambda)e_+(\lambda)]^{-1}O(\lambda^{-2}) & [\hat{e}(\lambda)e_+(\lambda)]^{-1}O(\lambda^{-2}) & e_+(\lambda)^{-2}(1-r_{33})\hat{\omega}_{33}(\lambda), \end{bmatrix} \tag{7.2}$$

which is

$$\mathbb{R}_l(\lambda) - \mathbb{R}_r(\lambda) =$$

$$
\left[
\begin{array}{cc}
i\hat{\omega}_{11}(\lambda) - e(\lambda)^{-2}\omega(\lambda)_{11} & (i-1)\hat{\omega}_{12}(\lambda) - [e(\lambda)\hat{e}(\lambda)]^{-1}O(\lambda^{-1/2}) \\[2em]
-(i+1)\hat{\omega}_{21}(\lambda) - [e(\lambda)\hat{e}(\lambda)]^{-1}O(\lambda^{-1/2}) & -i\hat{\omega}_{22}(\lambda) - \hat{e}(\lambda)^{-2}\omega_{22}(\lambda) \\[2em]
O(\lambda^{-3}) - [e(\lambda)e_{+}(\lambda)]^{-1}O(\lambda^{-2}) & O(\lambda^{-3}) - [\hat{e}(\lambda)e_{+}(\lambda)]^{-1}O(\lambda^{-2})
\end{array}
\right.
$$

$$
\left.
\begin{array}{c}
\dfrac{Q}{P}(i-1)\lambda^{1/2}\hat{\omega}_{13}(\lambda) + [e(\lambda)e_{+}(\lambda)]^{-1}r_{13}\lambda^{3/2}\hat{\omega}_{13}(\lambda) \\[2em]
\dfrac{Q}{P}(1-i)\lambda^{1/2}\hat{\omega}_{23}(\lambda) + [\hat{e}(\lambda)e_{+}(\lambda)]^{-1}r_{23}\lambda^{3/2}\hat{\omega}_{23}(\lambda) \\[2em]
-1 + O(\lambda^{-2.5}) - e_{+}(\lambda)^{-2}(1-r_{33})\hat{\omega}_{33}(\lambda)
\end{array}
\right].
$$

$$(7.3)$$

We now recall that $\mathcal{L}$ is a dissipative operator, which means that its eigenvalues must be in the closed upper half–plane. So we may write

$$\lambda = \bar{x} + i\bar{y}, \quad \lambda^{1/2} = x + iy, \tag{7.4}$$

where $\bar{x} \in \mathbb{R}$, and $\bar{y}$, x, $y > 0$. Let us write the expressions for $e(\lambda)$, $\hat{e}(\lambda)$, and $e_{+}(\lambda)$ in terms of x and y. We recall definitions (6.1) and (4.83) to calculate $e(\lambda)$ as

$$e(\lambda) = e^{\gamma(\lambda)L} = e^{PL(x+iy)}(1 + O(\lambda^{-1/2})) = e^{PLx}e^{iPLy}(1 + O(\lambda^{-1/2})). \tag{7.5}$$

Notice that $e(\lambda)$ is unbounded in the upper half–plane, but $e(\lambda)^{-1}$ is bounded. In a similar manner, we calculate $\hat{e}(\lambda)$ as

$$\hat{e}(\lambda) = e^{i\hat{\gamma}(\lambda)L} = e^{iPL\lambda^{1/2}(1+O(\lambda^{-1}))} = e^{PLi(x+iy)}(1+O(\lambda^{-1/2})) = e^{-PLy}e^{iPLx}(1+O(\lambda^{-1/2})). \tag{7.6}$$

Notice that $\hat{e}(\lambda)$ is bounded in the upper half–plane, but $\hat{e}(\lambda)^{-1}$ is unbounded. We calculate $e_+(\lambda)$ as

$$e_+(\lambda) = e^{i\Gamma(\lambda)L} = e^{iQL\lambda(1+O(\lambda^{-2}))} = e^{-QL\bar{y}}e^{iQL\bar{x}}(1 + O(\lambda^{-1})). \qquad (7.7)$$

Notice that $e_+(\lambda)$ is bounded in the upper half–plane, but $e_+(\lambda)^{-1}$ is not. If we now multiply both sides of the reflection matrices by the non–singular matrix

$$\tilde{\mathbb{E}}(\lambda) = \begin{bmatrix} 1 & 0 & 0 \\ 0 & \hat{e}(\lambda) & 0 \\ 0 & 0 & e_+(\lambda) \end{bmatrix}, \qquad (7.8)$$

whose entries are bounded, then we will arrive at a matrix, whose entries are all bounded in the upper half–plane. Thus we have changed problem (7.1) into the one involving a matrix, whose entries are bounded in the upper half–plane. Thus, we are looking for the solutions of the following equation:

$$0 = \det(\mathbb{R}_l(\lambda) - \mathbb{R}_r(\lambda)) = \det \tilde{\mathbb{E}}(\lambda) \det(\mathbb{R}_l(\lambda) - \mathbb{R}_r(\lambda)) \det \tilde{\mathbb{E}}(\lambda)$$

$$= \det \left[\tilde{\mathbb{E}}(\lambda)(\mathbb{R}_l(\lambda) - \mathbb{R}_r(\lambda))\tilde{\mathbb{E}}(\lambda) \right]$$

$$= \det \begin{bmatrix}
i\hat{\omega}_{11}(\lambda) - e(\lambda)^{-2}\omega(\lambda)_{11} & (i-1)\hat{e}(\lambda)\hat{\omega}_{12}(\lambda) - e(\lambda)^{-1}O(\lambda^{-1/2}) \\[2mm]
-(i+1)\hat{e}(\lambda)\hat{\omega}_{21}(\lambda) - e(\lambda)^{-1}O(\lambda^{-1/2}) & -i\hat{e}(\lambda)^2\hat{\omega}_{22}(\lambda) - \omega_{22}(\lambda) \\[2mm]
e_+(\lambda)O(\lambda^{-3}) - e(\lambda)^{-1}O(\lambda^{-2}) & \hat{e}(\lambda)e_+(\lambda)O(\lambda^{-3}) - O(\lambda^{-2})
\end{bmatrix}$$

$$\begin{matrix}
\dfrac{Q}{P}(i-1)e_+(\lambda)\lambda^{1/2}\hat{\omega}_{13}(\lambda) + e(\lambda)^{-1}r_{13}\lambda^{3/2}\hat{\omega}_{13}(\lambda) \\[3mm]
\dfrac{Q}{P}(1-i)\hat{e}(\lambda)e_+(\lambda)\lambda^{1/2}\hat{\omega}_{23}(\lambda) + r_{23}\lambda^{3/2}\hat{\omega}_{23}(\lambda) \\[3mm]
-e_+(\lambda)^2 + e_+(\lambda)^2 O(\lambda^{-2.5}) - (1-r_{33})\hat{\omega}_{33}(\lambda)
\end{matrix} \Bigg] .$$

$$(7.9)$$

Let us expand this determinant with respect to the entries of the bottom row. Taking into account that $e_+(\lambda)$ and $e^{-1}(\lambda)$ are bounded in the upper half–plane, we can rewrite Eq.(7.9) as follows:

$$O(\lambda^{-1/2}) = [e_+(\lambda)^2 + (1 - r_{33})(1 + O(\lambda^{-1}))][(i - e(\lambda)^{-2})(-i\hat{e}(\lambda)^2 - 1)(1 + O(\lambda^{-1/2}))(\lambda)$$
$$+ \{(i+1)\hat{e}(\lambda)(1 + O(\lambda^{-1/2})) + e(\lambda)^{-1}O(\lambda^{-1/2})\}(i-1)\hat{e}(\lambda)(1 + O(\lambda^{-1}))].$$

$$(7.10)$$

7.2 The α–branch of the Spectrum

In this section, we derive the leading term of the asymptotic approximation for the α–branch of the spectrum. The estimate for the remainder term will be justified in Section 7.4. We now return to Eq.(7.10). Due to the fact that all terms in this equation are bounded, we can rewrite it in the following form:

$$\left[e_+^2(\lambda) + (1 - r_{33}) \right] \left[(e^{-2}(\lambda) - i)(i\hat{e}^2(\lambda) + 1) - 2\hat{e}^2(\lambda) \right] = O(\lambda^{-1/2}). \qquad (7.11)$$

58

If we replace the right–hand side of Eq.(7.11) with zero, we obtain a new equation, which we will call the *model equation*. In Sections 7.2 and 7.3, we will study the distribution of the roots of the model equation and in Section 7.4, we will use Rouche's Theorem to complete the proof of our main result.

We start with the equation

$$0 = e_+(\lambda)^2 + (1 - r_{33}). \tag{7.12}$$

We then substitute (7.7) and arrive at the equation

$$1 = r_{33} - e_+(\lambda)^2 = r_{33} - e^{i2LQ\lambda(1+O(\lambda^{-2}))}. \tag{7.13}$$

Let us consider the simpler equation

$$e^{i2LQ\lambda} = r_{33} - 1. \tag{7.14}$$

Obviously the solutions of Eq.(7.14) are given by the formula

$$i2LQ\mathring{\lambda}_n = \ln(r_{33} - 1) + 2\pi i n, \qquad n \in \mathbb{Z}. \tag{7.15}$$

Thus for each $n \in \mathbb{Z}$, we will have a corresponding λ_n. Solving for these λ_n, we have

$$\mathring{\lambda}_n^\alpha = -\frac{i}{2LQ} \ln(r_{33} - 1) + \frac{\pi n}{LQ}. \tag{7.16}$$

Next we substitute our expression for r_{33} found in (6.80) to have

$$\mathring{\lambda}_n^\alpha = \frac{i}{2LQ} \ln \left[\frac{g_\alpha + \sqrt{I_\alpha GJ}}{g_\alpha - \sqrt{I_\alpha GJ}} \right] + \frac{\pi n}{LQ}. \tag{7.17}$$

Formula (7.17) gives us the leading term in the asymptotical representation for the α–branch of the eigenvalues. Note that under our assumption $g_\alpha \neq \sqrt{I_\alpha GJ}$, the logarithmic term is well–defined.

59

7.3 The h–branches of the Spectrum

In this section, we investigate the roots of the second part of the model equation, i.e., we consider the following equation

$$0 = -\hat{e}(\lambda)^2 - i + ie(\lambda)^{-2}\hat{e}(\lambda)^2 - e(\lambda)^{-2}.$$ (7.18)

Using (7.5) and (7.6) and omitting the lower order terms, we write the latter equation in terms of x and y as

$$i = -e^{-2PLy}e^{2PLix} + ie^{-2PLx}e^{-i2PLy}e^{-2PLy}e^{2PLix} + e^{-2PLx}e^{-i2PLy}.$$ (7.19)

From the general theory, we know that the eigenvalues accumulate at infinity. There are three ways for λ to go to infinity. The simplest way is when $x = y \to \infty$. In that case, the right hand side of Eq.(7.19) would approach zero, while the left hand side is a constant. Due to the obvious contradiction, we exclude this case. We will investigate the two remaining cases. As the first case, we consider the domain for (x, y) bounded by the positive real semi–axis $0 \le x < \infty$ and the diagonal $x = y > 0$. As the second case, we consider the domain for $(x.y)$ bounded by the diagonal $x = y > 0$ and the positive imaginary semi–axis $0 \le y < \infty$.

Case 1. Let $x > y$, $x \to \infty$. We notice then that the last two terms in (7.18) approach zero. Discarding these lower order terms, we have a new model equation

$$-i = e^{-2PLy}e^{i2PLx}.$$ (7.20)

We are interested in finding those pairs (x, y) that will solve this equation. Eq.(7.20) can be rewritten in the form

$$e^{i(-\pi/2+2\pi n)} = e^{-2PLy+i2PLx}.$$ (7.21)

So we may write that

$$i(-\pi/2 + 2\pi n) = -2PLy_n + i2PLx_n.$$ (7.22)

Equating real and imaginary parts, we find that $y = 0$ and

$$-\frac{\pi}{2} + 2\pi n = 2PLx_n, \qquad n \ge 1,$$ (7.23)

60

or

$$x_n = \frac{\pi}{PL}(n - 1/4), \quad y_n = 0. \tag{7.24}$$

Thus we may write by comparison with (7.3) that

$$\overset{\circ}{\lambda}_n^{1/2} = \frac{\pi}{PL}(n - 1/4), \tag{7.25}$$

which we square to obtain an approximation to part of this branch of the eigenvalues

$$\overset{\circ}{\lambda}_n^h = \left(\frac{\pi}{PL}\right)^2 (n - 1/4)^2, \qquad n \geq 1. \tag{7.26}$$

Formulae (7.26) gives us the leading term in the asymptotics. The lower order terms will be discovered in Section 7.3.

Case II. Let $y > x$, $y \to \infty$. Discarding lower order terms, we have the new model equation from (7.19)

$$i = e^{-2PLx}e^{-i2PLy}. \tag{7.27}$$

The equivalent form of Eq.(7.27) is

$$e^{i(\pi/2 + 2\pi n)} = e^{-2PLx_n - i2PLy_n}. \tag{7.28}$$

So we may write that

$$i\left(\frac{\pi}{2} + 2\pi n\right) = -2PLx_n - i2PLy_n. \tag{7.29}$$

Upon equating real and imaginary parts, we obtain

$$y_n = \frac{\pi}{PL}(n - 1/4), \quad x_n = 0. \tag{7.30}$$

Thus, we may write by comparison with (7.4) that

$$\overset{\circ}{\lambda}_n^{1/2} = i\frac{\pi}{PL}(|n| - 1/4), \qquad n \leq -1, \tag{7.31}$$

which we square to obtain an approximation to the second part of this branch of the eigenvalues as

$$\overset{\circ}{\lambda}_n^h = -\left(\frac{\pi}{PL}\right)^2 (|n| - 1/4)^2. \tag{7.32}$$

The lower order asymptotical terms will be discussed in Section 7.4.

61

7.4 Rouche's Theorem

In this section, we justify the accuracy of the calculated eigenvalues. Our main tool is Rouche's Theorem. By using this theorem, we will find circles of radii ϵ_n around the approximated eigenvalues $\mathring{\lambda}_n$ displayed in (7.19), (7.26), and (7.32) so that the actual eigenvalue λ_n is in the circle of radius ϵ_n centered at $\mathring{\lambda}_n$.

Returning to Eq.(7.11), we rewrite it in the form

$$K_1(\lambda)K_2(\lambda) = O(\lambda^{-1/2}), \tag{7.33}$$

where complex–valued functions K_1 and K_2 are defined by the formulae

$$K_1(\lambda) = e_+^2(\lambda) + (1 - r_{33}), \quad K_2(\lambda) = \hat{e}^2(\lambda) - i + ie^{-2}(\lambda)\hat{e}^2(\lambda) - e^2(\lambda), \tag{7.34}$$

with r_{33} being defined in (6.60). We present a very detailed proof for the case of the α–branch eigenvalues. All proofs for the h–branch eigenvalues can be done in a similar fashion. Let us consider a sufficiently distant eigenvalue λ_n^α from the α–branch and show that there exists a circle of a small radius ϵ_n centered at the root of Eq.(7.16), which we denoted as $\mathring{\lambda}_n^\alpha$, such that λ_n^α is exactly inside the circle. For the aforementioned circle, we will use the notation $B_{\epsilon_n}(\mathring{\lambda}_n^\alpha)$. It can be directly verified that the following estimate is valid:

$$\sup_{n \in \mathbb{Z}} |K_2(\mathring{\lambda}_n^\alpha)| < \infty. \tag{7.35}$$

We will proceed with the proof assuming that the branches are separated, i.e.,

$$\inf_{n,m \in \mathbb{Z}} |\mathring{\lambda}_n^\alpha - \mathring{\lambda}_m^h| = d > 0. \tag{7.36}$$

We note that if condition (7.36) is not satisfied, the proofs would be technically more complicated. However, the main direction of the proof will be essentially the same. Taking into account (7.36), we complement (7.35) with an additional estimate

$$\inf_{n \in \mathbb{Z}} |K_2(\mathring{\lambda}_n^\alpha)| > 0. \tag{7.37}$$

Using (7.35) and (7.37), we can rewrite Eq.(7.33) for $\lambda \in B_{\epsilon_n}(\mathring{\lambda}_n^\alpha)$ in the form

$$K_1(\lambda) = O(\lambda^{-1/2}). \tag{7.38}$$

In what follows, we will use the following version of Rouche's Theorem:

Assume that f and g are two analytic functions in the closed disk centered at the point a of radius r (which we denote as $B_r(a) \cup \partial B_r(a)$). If the following estimate is valid

$$|f(\lambda)| > |g(\lambda)| > 0, \quad \lambda \in \partial B_r(a), \tag{7.39}$$

then the number of zeros counting their multiplicities of the functions $(f + g)$ and f coincide in $B_r(a)$.

Recalling from (6.1) the definition of $e_+(\lambda)$, we may rewrite Eq.(7.34) as

$$e^{i2\Gamma(\lambda)L} + (1 - r_{33}) = O(\lambda^{-1/2}). \tag{7.40}$$

Substituting the expression for $\Gamma(\cdot)$ from (4.85), we have

$$\begin{aligned}
&e^{i2L(Q\lambda+O(\lambda^{-1}))} + (1 - r_{33}) + O(\lambda^{-1/2}) = \\
&e^{i2LQ\lambda}e^{O(\lambda^{-1})} + (1 - r_{33}) + O(\lambda^{-1/2}) = 0.
\end{aligned} \tag{7.41}$$

Simplifying Eq.(7.41), we arrive at the desired form

$$e^{i2LQ\lambda} + (1 - r_{33}) = O(\lambda^{-1/2}). \tag{7.42}$$

It is convenient to introduce a new function

$$\hat{K}_1(\lambda) = e^{i2LQ\lambda} + (1 - r_{33}). \tag{7.43}$$

and rewrite Eq.(7.41) in the form

$$\hat{K}_1(\lambda) + O(\lambda^{-1/2}) = 0. \tag{7.44}$$

Let $g(\lambda)$ be an analytic function defined by

$$g(\lambda) = K_1(\lambda) - \hat{K}_1(\lambda). \tag{7.45}$$

We evaluate the function $\hat{K}_1(\lambda)$ on the circle of radius ϵ_n about λ_n° from (7.17). The the estimate of the value of ϵ_n will be identified later. We have

$$\begin{aligned}
\hat{K}_1(\lambda_n^\circ + \epsilon_n e^{i\varphi}) &= e^{i2LQ(\lambda_n^\circ + \epsilon_n e^{i\varphi})} + (1 - r_{33}) \\
&= e^{i2LQ\lambda_n^\circ}e^{i2LQ\epsilon_n e^{i\varphi}} + (1 - r_{33}), \qquad 0 \leq \varphi < 2\pi.
\end{aligned} \tag{7.46}$$

Recalling that λ_n° is a solution of the equation $\hat{K}_1(\lambda) = 0$, we replace the first exponential with $(1 - r_{33})$ and factor it out to obtain

$$\hat{K}_1(\lambda_n^\circ + \epsilon_n e^{i\varphi}) = (1 - r_{33}) \left[e^{i2LQ\epsilon_n e^{i\varphi}} - 1 \right], \qquad 0 \le \varphi < 2\pi. \tag{7.47}$$

We apply the Mean Value Theorem and have

$$
\begin{aligned}
\hat{K}_1(\lambda_n^\circ + \epsilon_n e^{i\varphi}) &= (1 - r_{33}) \left. \frac{d}{d\epsilon_n} e^{i2LQ\epsilon_n e^{i\varphi}} \right|_{\epsilon_n = \xi_n} \epsilon_n \\
&= (1 - r_{33}) i2LQ e^{i\varphi} e^{i2LQ\xi_n e^{i\varphi}} \epsilon_n. \qquad 0 < \xi_n < \epsilon_n.
\end{aligned}
\tag{7.48}
$$

From (7.48), we immediately obtain the estimate

$$\left| \hat{K}_1(\lambda_n^\circ + \epsilon_n e^{i\varphi}) \right| = |1 - r_{33}| \, |2LQ| \, |1 + O(\xi_n)| \, |\epsilon_n|, \qquad 0 < \xi_n < \epsilon_n. \tag{7.49}$$

It is clear that for a large enough n, we can bound the term $|1 + O(\xi_n)| > C_0$ for $0 < C_0 < 1$. For convenience of further calculation, we take $C_0 = 1/2$. Thus for large enough n, we have

$$\left| \hat{K}_1(\lambda_n^\circ + \epsilon_n e^{i\varphi}) \right| > |1 - r_{33}| \, |LQ| \, |\epsilon_n|. \tag{7.50}$$

Now we turn to $g(\lambda)$ as defined in (7.45). Evaluating this function on the circle $\lambda = \lambda_n^\circ + \epsilon_n e^{i\varphi}$, $0 \le \varphi < 2\pi$ and then using formula (7.17) for λ_n° we calculate that

$$\left| g(\lambda_n^\circ + \epsilon_n e^{i\varphi}) \right| = \left| O((\lambda_n^\circ + \epsilon_n e^{i\varphi})^{-1/2}) \right| = \left| O(n^{-1/2}) \right| < \frac{C}{\sqrt{n}} \tag{7.51}$$

for some absolute constant C. In order to apply Rouche's Theorem, we would like to show that

$$\left| \hat{K}_1(\lambda_n^\circ + \epsilon_n e^{i\varphi}) \right| > \left| g((\lambda_n^\circ + \epsilon_n e^{i\varphi})^{-1/2}) \right|, \qquad 0 \le \varphi < 2\pi. \tag{7.52}$$

Taking into account (7.50), relation (7.51) can be rewritten as

$$\left| \hat{K}_1(\lambda_n^\circ + \epsilon_n e^{i\varphi}) \right| > |1 - r_{33}| \, |LQ| \, |\epsilon_n| \ge \frac{C}{\sqrt{n}} > \left| g(\lambda_n^\circ + \epsilon_n e^{i\varphi}) \right|. \tag{7.53}$$

If we choose a circle of radius $\epsilon_n = \tilde{C}/\sqrt{n}$, where

$$\tilde{C} = \frac{C}{|1 - r_{33}| \, |LQ|}, \tag{7.54}$$

64

then we will be able to apply Rouche's Theorem. The latter fact justifies the accuracy of the asymptotic formulae (3.3).

As was already mentioned, the justification of the h–branch spectral asymptotics can be done in a similar fashion.

CHAPTER VIII
Conclusion

In the present paper, we investigated asymptotic properties of a system of two coupled differential equations in two unknown functions h and α. This system was supplied with a two parameter family of nonselfadjoint boundary conditions, which have been introduced to model the action of smart materials on a beam. We have found a related Dynamics Generator, which governs the vibrations of the model and denoted it by $\mathcal{L}$. We formulated the important properties of this operator, and commenced to compute the asymptotics of it's spectrum.

To compute the spectral asymptotics, we found an operator pencil related to the operator $\mathcal{L}$. This operator pencil has the same spectrum as the operator $\mathcal{L}$. To find the spectral asymptotics for the pencil, we have solved a highly nonstandard boundary–value problem, which consists of the sixth order ordinary differential equation and six boundary conditions. Both the equation and the three boundary conditions contain the spectral parameter. To find a fundamental system of solutions of the aforementioned ordinary differential equation, we have analyzed the characteristic equation. We have used Cardano's Formulae to compute the approximations for the six solutions to the characteristic equation related to the Operator Pencil's spectral equation. Having six roots of the characteristic equation, we have constructed the general solution to the spectral equation of the operator pencil. To apply the boundary conditions of the operator pencil, we have developed *a new method*. Namely, we have introduced the *left* and *right reflection matrices*, which allowed us to simplify the problem and make an analytical study possible. In numerous papers on this bending–torsion vibration model, the only study performed by different authors were either numerical simulations or wind tunnel experimentations.

We have found two branches of the spectrum, and computed the spectral asymptotics of each branch by making use of the well known Rouche's Theorem. One branch lies asymptotically close to a line in the upper half-plane and parallel to the horizontal axis. The other branch lies in the upper half-plane and is close to the horizontal axis. These calculations give the approximation to the spectra of both the operator pencil, and of the operator $\mathcal{L}$. The geometry of the spectrum tells us that solutions to the original system of coupled differential equations though stable, are not uniformly stable.

Our future research plans contain the following steps:

- Calculation of asymptotic representations for the eigenfunctions of the operator $\mathcal{L}_{g_h g_\alpha}$; the asymptotics is assumed to be with respect to the number of the eigenvalue, when this number tends to infinity and is assumed to be uniform with respect to the spatial variable $x \in [-L, 0]$;

- Proof that the set of eigenfunctions is complete in the energy space;

- Proof that the set of eigenfunctions is minimal (linearly independent) in the energy space;

- Study the cases when the set of eigenfunctions forms an unconditional basis (the Riesz basis) of the energy space;

- Apply the asymptotic, spectral, and the Riesz basis property results to solve different boundary and distributed control problems using the method of spectral decomposition.

Acknowledgement.

The first author is grateful to the Flight Systems Research Center of the University of California at Los Angeles and especially to its Director, Prof. A.V. Balakrishnan, for the opportunity to spend a one–year period at FSRC under the support of the National Science Foundation grant DMS # 9972748 (IGMS). Partial support by the National Science Foundation Grants ECS #0080441, DMS#0072247, and the Advanced Research Program-2002 of Texas Grant 0036-44-016 is highly appreciated by the first author as well. Partial support by two NASA/Texas Space Grant Consortium Fellowships is highly appreciated by the second author.

BIBLIOGRAPHY

[1] Ashley, H. and Landahl, M., *Aerodynamics of Wings and Bodies*, Dover Publ. Inc., New York, (1985).

[2] Bisplinghoff, R.L., Ashley, H., and Halfman, R.L., *Aeroelasticity*, Dover Publ. Inc., New York, (1996).

[3] Fung, Y.C., *An Introduction to the Theory of Aeroelasticity*, Dover Publ. Inc., New York, (1993).

[4] Balakrishnan, A.V., Control of structures with self-straining actuators: coupled Euler/ Timoshenko model, *Nonlinear Problems in Aviation and Aerospace*, Gordon and Breach Sci. Publ., Reading, United Kingdom, (1998).

[5] Balakrishnan, A.V., Damping performance of strain actuated beams, *Comput. Appl. Math.*, **18**, (1), (1999), p.31–86.

[6] Balakrishnan, A.V., Theoretical limits of damping attainable by smart beams with rate feedback. In: *Smart Structures and Materials, 1997; Mathematics and Control in Smart Structures*, Vasundara V., Jagdish C., Eds., Proc. SPIE, **3039**, (1997), p.204–215.

[7] Shubov, M.A., Spectral operators generated by Timoshenko beam model, *System and Control Letters*, **38**, (4–5), (1999), p.249–258.

[8] Shubov, M.A., Exact controllability of damped Timoshenko beam, *IMA J. of Mathematical Control and Inf.*, **17**, (2000), p.375–395.

[9] Shubov, M.A., Asymptotics and spectral analysis of Timoshenko beam model, *Mathematische Nachrischten*, **241**, (2002), p.125–162.

[10] Balakrishan, A.V., Aeroelastic control with self–straining actuators: continuum models. in: *Smart Structures and Materials, Mathematics Control in Smart Structures*, Varadan V., Ed., *Proceedings of SPIE*, **3323**, (1998), p.44–54.

[11] Balakrishnan, A.V., Subsonic flutter suppression using self-straining acutators, to appear in: *Journal of the Franklin Institute*, Special Issue on Control, Udwadia F.,Ed., **338**, (2/3), (2001), p.149–170.

[12] Tzou, H.S., and Gadre, M., Theoretical analysis of a multi-layered thin shell coupled with piezoelectric shell actuators for distributed vibration controls, *J. Sound Vibr.*, **132**, p.433–450, (1989).

[13] Yang, S.M., and Lee, Y.J., Modal analysis of stepped beams with piezoelectric materials, *J. Sound Vibr.*, **176**, p.289–300.

[14] Balakrishnan, A.V., Shubov, M.A., Peterson, C.A., and Martin, C.A., Asymptotic and Spectral results for nonselfadjoint operators generated by coupled Euler–Bernoulli and Timoshenko beam model, Preprint of Texas Tech University, (2002).

[15] Adams, R.A., *Sobolev Spaces*, Academic Press, New York, (1975).

[16] Sz.–Nagy, B., and Foias, C., *Harmonic Analysis of Operators on Hilbert Space,* North Holland Publ. Co., (1970).

[17] Gohberg, I.Ts., and Krein, M.G., *Introduction to the Theory of Linear Nonselfadjoint Operators*, Trans. of Math. Monogr., **18**, AMS, Providence, RI, (1996).

[18] Istratescu, V.I., *Introduction to Linear Operator Theory*, Pure Appl. Math Series of Monog., Marcel Dekker Inc., New York, (1981).

[19] Marcus, A.S., *Introduction to the Spectral Theory of Polynomial Operator Pencils*, Transl. Math. Monogr., AMS, Vol. **71**, (1988).

[20] Artin, M., *Algebra*, Prentice-Hall, Inc., New Jersey, (1991).

REPORT DOCUMENTATION PAGE

1. AGENCY USE ONLY (Leave blank)	2. REPORT DATE	3. REPORT TYPE AND DATES COVERED
	November 2003	Contractor Report

4. TITLE AND SUBTITLE

Asymptotic Distribution of Eigenfrequencies for a Coupled Euler-Bernoulli and Timoshenko Beam Model

5. FUNDING NUMBERS

UCLA Grant Number NCC4-121

6. AUTHOR(S)

Marianna A. Shubov and Cheryl A. Peterson

7. PERFORMING ORGANIZATION NAME(S) AND ADDRESS(ES)

NASA Dryden Flight Research Center
P.O. Box 273
Edwards, California 93523-0273

8. PERFORMING ORGANIZATION REPORT NUMBER

H-2528

9. SPONSORING/MONITORING AGENCY NAME(S) AND ADDRESS(ES)

National Aeronautics and Space Administration
Washington, DC 20546-0001

10. SPONSORING/MONITORING AGENCY REPORT NUMBER

NASA/CR-2003-212022

11. SUPPLEMENTARY NOTES

UCLA Grant Number NCC4-121; NASA Technical Monitor Kenneth W. Iliff, NASA Dryden Flight Research Center. Other contributors were the National Science Foundation (Division of Mathematical Science and Division of Engineering Sciences), the Advanced Research Program-2002 of Texas, and partial support by two NASA/Texas Space Grant Consortium Fellowships.

12a. DISTRIBUTION/AVAILABILITY STATEMENT

Unclassified—Unlimited
Subject Categories 02, 59, 66

This report is available at http://www.dfrc.nasa.gov/DTRS/

12b. DISTRIBUTION CODE

13. ABSTRACT (Maximum 200 words)

This research is devoted to the asymptotic and spectral analysis of a coupled Euler-Bernoulli and Timoshenko beam model. The model is governed by a system of two coupled differential equations and a two parameter family of boundary conditions modelling the action of self-straining actuators. The aforementioned equations of motion together with a two-parameter family of boundary conditions form a coupled linear hyperbolic system, which is equivalent to a single operator evolution equation in the energy space. That equation defines a semigroup of bounded operators. The dynamics generator of the semigroup is our main object of interest. For eash set of boundary parameters, the dynamics generator has a compact inverse. If both boundary parameters are not purely imaginary numbers, then the dynamics generator is a nonselfadjoint operator in the energy space. We calculate the spectral asymptotics of the dynamics generator. We find that the spectrum lies in a strip parallel to the horizontal axis, and is asymptotically close to the horizontal axis. - thus the system is stable, but is not uniformly stable.

14. SUBJECT TERMS

Aeroelastic mode, Bending-torsion vibration model, Integro-differential equations, Reflection matrix, Theodorsen function

15. NUMBER OF PAGES

79

16. PRICE CODE

17. SECURITY CLASSIFICATION OF REPORT	18. SECURITY CLASSIFICATION OF THIS PAGE	19. SECURITY CLASSIFICATION OF ABSTRACT	20. LIMITATION OF ABSTRACT
Unclassified	Unclassified	Unclassified	Unlimited

www.ingramcontent.com/pod-product-compliance
Lightning Source LLC
Chambersburg PA
CBHW081241180526
45171CB00005B/497